Horizontal and Vertical Drilling

Byron Davenport

McGraw-Hill, Inc.

New York St. Louis San Francisco Auckland Bogotá
Caracas Lisbon London Madrid Mexico Milan
Montreal New Delhi Paris San Juan São Paulo
Singapore Sydney Tokyo Toronto

Library of Congress Cataloging-in-Publication Data

Davenport, Byron.
    Horizontal and vertical drilling / Byron Davenport.
        p.   cm.
    Includes index.
    ISBN 0-07-015491-0
    1. Horizontal oil well drilling.   2. Oil well drilling.
    I. Title.
    TN871.25.D38   1992
    622'.3382—dc20   91-46034
                                                        CIP

1 2 3 4 5 6 7 8 9 0   DOC/DOC   9 8 7 6 5 4 3 2

ISBN 0-07-015491-0

*The sponsoring editor for this book was Gail F. Nalven, the editing supervisor
was Jane Palmieri, and the production supervisor was Pamela Pelton. It was
set in Old Times Roman by Techna Type, Inc., York, Pennsylvania.*

Printed and bound by R. R. Donnelley & Sons Company.

*Portions of this book were previously published under the title Handbook of
Drilling Practices.*

Contents

Preface vii

Dedication and Acknowledgments ix

1
A Typical Oil and Gas Transaction **1**

2
The AFE **13**

3
Drilling Contracts **28**

4
The Drilling Prognosis **35**

5
Key Rental Items **47**

6
Rig-up, Spud-in, and Setting
Surface Casing **57**

7
Nippling Up the BOP Stack **75**

8
Setting Up the Bottom Hole Assembly **84**

9
Drilling Out Surface Casing, Tests,
and Squeeze Jobs **97**

10
Drill Bits **111**

11
The Mud Program **118**

12
Drilling Ahead **124**

13
Produce-While-Drilling (PWD) Equipment **127**

14
Key Maintenance **134**

15
Special Problems During Vertical Drilling **138**

16
Fishing Tools **161**

17
Drill Stem Tests **175**

18
Coring the Well **184**

19
*Logging the Well and
Accompanying Problems* **189**

20
*The Intermediate String, Liners,
and Testing* **194**

21
Finding the Horizontal Zone to Drill **203**

22
*The Horizontal Directional Driller and MWD
Tools* **206**

23
Drilling the Curve **209**

24
Special Problems During Horizontal Drilling .. **214**

25
Setting the Packer in a Horizontal Well **227**

26
The Long String and the Cement Job **229**

27
Finishing the Well and the Paperwork **238**

28
Plug-and-Abandon Procedures **242**

29
Blowout Control and Calculations **247**

30
Oilfield Firefighting **266**

31
Completing a Horizontal Well **275**

Appendix A
IADC Footage Drilling Contract **279**

Appendix B
Common Oilwell Drilling Calculations **305**

Appendix C
The Cement Book **314**

Appendix D
The Consultant's Checklist **319**

Appendix E
Capacity and Displacement of Drill Collars ... **328**

Appendix F
Pump Output Table **330**

Glossary **333**

Index **349**

Preface

This book's purpose is to describe the technical as well as nontechnical efforts of drilling vertical and horizontal wells, so that engineers, consultants, salespersons, roughnecks, and the general public may read and understand what actually happens in the field.

Many, many books have been written on petroleum engineering and drilling techniques for vertical wells, most of which the average oilfield hand cannot understand because of the high-level mathematics that means nothing in the field. Of course, all the office engineers will disagree with that statement; however, in the field everything is much different. Very few books have been written on horizontal drilling and actual field problems and practices. I realized that much of what I learned in the books did not apply to actual field operations. Horizontal wells are "seat-of-the-pants" operations that can be drilled in a safe or dangerous manner. You cannot drill a horizontal well with a calculator or any books I have reviewed. New terminology in horizontal wells has evolved, shaking up the average older oilfield hand.

Engineers from the office would occasionally come out to look over an operation in the field. It became very clear to me that they did not have a handle on what they were doing. Writing a well design, casing designs, and AFE's is much different when you get in the field. For years I would personally train and supervise young engineers who would come to the field unaware of what they were actually seeing.

This book is designed to give engineers an insight into the reality of rig operations and at the same time to give roughnecks, drillers, and toolpushers a better understanding of how engineers think.

For the service hands or the company representatives, the book explains just exactly how oil wells are drilled vertical and horizontal and will make them more knowledgeable when they go into the field trying to sell their products. I think I share with all other consultants the irritation of having a salesperson come into the field not knowing the oil business. So, if you are a salesperson, this book should give you the understanding of how to drill a well so you can carry on a good conversation with a consultant or engineer on location.

This book is presented not as the last word in drilling, since the technology changes daily, but simply as a guide and reference manual to solve some major problems encountered while drilling oil or gas wells. I hope it will expose some of the hazards and hardships that it takes to become a professional oilman. Most good oilmen are willing to stay on the floor for two or three days if necessary. They stay to solve major problems at a moment's notice while everyone else is sleeping.

The book has been written so that each problem can be referenced quickly. Although the publisher and I take no liability if some of the provided methods do not work, everything discussed in this manual has worked better in most cases than other methods. Also, the advice in this manual has worked in drilling oil wells in and around the Gulf Coast, and if you can drill on the Gulf Coast of the United States, you can drill anywhere in the world. Each area has its own peculiarities, but this manual covers the basics that apply to a land rig anywhere in the world.

Byron "Duke" Davenport

Dedication and Acknowledgments

This book is dedicated to Robert Lindal Davis, known to everyone as Bob Davis. My late grandfather was a pioneer in oilwell firefighting, retiring after an accident, only to become a giant in the petrochemical industry on the Houston ship channel. He was the strength of our whole family and was a man to look up to. He told me in his last years that he had outlived all his friends, but being oilfield hands, they were probably still hard at work in heaven. My late grandmother Ruby Alma Davis was his helpmate for 56 years and a dedicated family lady. Both of them encouraged my entry into the oilfield. They were both God-fearing people who taught me a lot about life. God bless them both.

I would like to thank my children, I'Onika, Byron, Robert, Melissa, David ''Little Duke,'' and Paul, for their understanding during the many hours Dad was gone fighting oilwell fires or drilling wells only to return, sometimes after months of absence, for two or three days and to be called out again.

Thanks also to the many friends I have made in the patch. I've always been a good listener and learned many tricks of the trade while listening to old-timers tell their stories in coffee shops. Some of the oilfield greats who have been lifetime friends are Bob Petifils, Al Thomas, Ed Moses, Red Garrett, Jack Bradford, Gene Pickett, Herb Alexander, Harold Sadler, Red

Adair, Myron McKinley, Howard Grantham, Tom
Largé, Charlie Duncan, Louis Trojack, David Nix,
Jerry Trythall, Jim Nelson, Nelson Bunker Hunt,
Sammy Magolito, Vicente Barrerra, ''Doc'' Davis,
Robert Scoggins, and last but not least George Khoury,
all of whom taught me a lot about different well
problems and situations. TYJ

1
A Typical Oil and Gas Transaction

Before an oil or gas well can be drilled and even before the consultant or engineer ever sees the prospect many things have to be done. The following is a typical structure of an oil and gas drilling transaction.

The principal party in an oil well drilling operation is the *operator*. This is the "oil company," either a well-known major company or an independent. The Texas Railroad Commission (which regulates all drilling and production in Texas) officially designates this party as the operator, and we will use that term. Every well in Texas should have a sign on it designating the operator and the name of the well. The operator employs the *drilling consultant*, who is there to protect the operator's interest.

The operator also hires a *geologist*, who locates an area that he feels is a good prospect for petroleum. The geologist may recommend a *development well* that attempts to hit a structure already producing or he may recommend drilling a *wildcat well* into an untested structure.

1

The operator next directs a *landman* to acquire drilling rights. A major company will usually have a paid staff of geologists and landmen. A small independent may engage outsiders for a fee. The landman determines who owns the mineral rights in the area to be drilled and then attempts to acquire lease rights from the *landowner* through a document called the *oil and gas lease*. He will generally have to negotiate terms that are acceptable to the operator, because a lease that is too burdensome to the operator will kill the deal and cause the operator to look elsewhere.

The items the landman must negotiate include:

- Bonus
- Delay rentals
- Royalties
- Length of time within which drilling must commence
- Payment for and/or restrictions against surface damage and water use, etc.

The landman pays the lease bonus to the landowner when the lease is signed, normally with a draft payable in 30 to 60 days. The landman will then use that period before the draft is paid to examine the title. If he is not satisfied with the title, he cancels the draft, and the lease is cancelled.

To check the title the landman obtains from the landowner any *abstract of title* from an *abstract company,* or he may sometimes check the title directly from the county records. He may have to do certain curative work to correct any defects in the title. For instance, there may be a prior oil lease outstanding on the property, which needs to be cleared from the record. If the landman does not do his job properly the operator may find that oil refineries and gas pipelines are unwilling to purchase the well output because he cannot show that he owns it. The operator may spend a lot of money to drill a well and have someone else walk up and say, ''Thank you for my new well.''

The operator then contacts one or more *investors*. These may be other oil companies, but they are usually outsiders who wish to invest in an oil well but lack the expertise to do so directly. They will entrust their money to the operator in exchange for part ownership of the well. The operator will typically sell about 75% of his interest in the well to the investors who will pay 100% of the drilling costs. The operator will thus get about a 25% interest, sometimes called a ''carried interest'' in the well, at no cost, in exchange for his efforts and his skill.

The operator then hires the *drilling contractor* who owns the drilling rig and employs the crew to drill the well. *A drilling contract,* which sets out in detail the obligations of the contractor, is signed by both parties. The consultant at the drill site should be familiar with this contract. It will determine which services and equipment will be provided by the contractor for his fee and which are extras to be provided by the operator. The operator also hires specialists to perform other services such as casing, cementing, logging, perforating, fracturing, acidizing, lost tool recovery, drilling fluid preparations, etc. The geologist is used again to analyze the drilling results and to determine which zones, if any, are worth producing. If there are one or more good zones, the well will be *completed* for production. If there is more than one good zone, there may be a multiple completion for producing several zones simultaneously. If there are no good zones, the well will be *plugged and abandoned* in accordance with the regulations that protect the water zones drilled through. The operator cannot just pick up the rig and leave an open hole.

The operator is then responsible for producing and selling the petroleum. The actual on-site production is handled by a *pumper,* either an employee or an independent serviceman, who has a route of wells that he visits periodically.

The landman is again engaged to:

1. Prove to the purchasing refinery or pipeline that the operator has good title to the output.
2. Determine how the proceeds are to be distributed.

Basic Terms of an Oil and Gas Lease

Almost all oil and gas leases follow the same basic format. The oil and gas lease cannot be compared to an ordinary apartment or real estate lease because they are not at all alike. Basically, the oil and gas lease permits the operator the right to explore for and produce petroleum. The landowner's rights consist mostly of the right to receive money. He receives three kinds of payments:

1. A bonus
2. Delay rentals
3. Royalties

The landowner is paid a one-time payment of a *bonus* when he signs the lease, which he keeps whether a well is ever drilled. In exchange for the bonus the operator has the right for a limited period called *the primary term* to drill for petroleum. If the primary term is for more than one year and the operator has not yet drilled a producing well, he must pay at the end of each year of the primary term a *delay rental* to extend the lease for another year. The lease may not be extended beyond the primary term by the payment of delay rentals, but if a successful well is drilled the lease is automatically extended so long as oil or gas is produced. In other words, when a successful well is completed the lease will last for as long as that well or any other well produces, and no delay rentals, which are a payment for the privilege of delaying drilling must be paid.

Once the well is producing, the landowner is entitled to a *royalty* of a certain percentage of production. In the past, one-eighth of production has been the standard royalty. This royalty is paid free and clear of all costs of drilling and all costs of production. The operator and his investors then own the other seven-eighths of production. This percentage is called the *working interest*. The working interest differs from the royalty in that:

1. It must pay all costs of drilling and all costs of production even though it is composed of only seven-eighths of the revenue.
2. Those that receive it control all drilling and production decisions.

If the operator has sold 75% of his interest to outside investors, then what he has sold is 75% of the seven-eighths working interest, and he has retained 25% of the seven-eighths working interest. The royalty holder has only the passive right to payment after the oil is produced. You should be aware that the one-eighth royalty is traditional, but that amount, like all terms of the lease, is negotiable and may be higher if the land is highly desirable (but is rarely lower than one-eighth).

The lease contains many provisions that may be of concern to the consultant, including a description of the surface area, the drilling location, surface damages, and perhaps specifications for restoring the surface. The lease may specify or restrict the use of water found on the premises for drilling mud, and it may limit access to the drill site to certain roads or entrances. The lease may even require the operator to build a road at his expense that becomes the property of the landowner. Most importantly, it may provide for early termination of the lease if drilling is not begun by a certain date, making the drilling schedule highly critical.

Petroleum Land Titles

This book is not intended to make you into either a surveyor or landman, but some familiarity with the principles of petroleum land titles is useful to anyone in the business. After oil is removed from the ground it is sold and dealt with just like any other item of personal property, but while still in the ground it is governed by the rules governing real estate transactions. In the United States, the surface owner generally has title to all air rights over the land and all minerals under the land, including oil and gas. In some foreign countries the government retains title to minerals, and the landowner has only surface rights. Oil and gas are also subject to a unique characteristic not applicable to other minerals—they can travel without regard for legal boundaries.

A landowner owns all oil produced on his land even if the reservoir extends under another person's land and he is draining oil from the entire reservoir. Of course, the well must reach bottom on his own land, and he cannot slant or directional drill to reach bottom on someone else's property. Thus, a landowner is subject to being deprived of his oil if his reservoir extends as far as a well on someone else's property. The principal defense to this is to drill an ''offset well'' to recover one's own oil or gas before it is lost (''pooling'' and ''unitization'' are other answers, but they are beyond the scope of this discussion). The horizontal well has changed the way people look at land boundaries. Now the boundaries are regulated in many states; however, in some states they are not. The directional driller's job is to not exceed the legal boundary. He does this by giving accurate directional surveys to the consultant on location and by following a preset directional chart. One foot could cause a legal problem if it were drilled into another person's boundary. A cement plug may be all that is necessary, but do not count on anything concerning regulation being simple. Politics plays a big part in drilling wells. Sometimes it comes down to how the regulator feels that day.

Ownership of land is determined by finding the original grant from the sovereign to a private owner and then tracing each transfer down to the present. Most states west of the Mississippi were originally federal territories with the land owned by the federal government. Titles in these states will generally start with a federal grant. Texas was never a federal territory, and all land titles in Texas trace back to a grant from the King of Spain, the Republic of Mexico, the Republic of Texas, or the State of Texas. The title then changes hands through various transfers, such as deeds, wills, intestate inheritances, tax sales, and mortgage foreclosures. A transfer may cover the entire property or it may include only a portion, such as the transfer of part or all of the minerals to one person, with surface rights left to another.

There are other documents that are not transfers but that do affect title such as mortgages, oil and gas leases, easements, etc. Almost all of these transfers are evidenced by a document that is recorded in the county where the land lies. The documents are a public record and thus give notice to anyone who seeks to determine title to a tract of property. Some transfers, however, such as an intestate inheritance (an inheritance by law from a person who left no will) or title obtained by adverse possession (title obtained by using the land and claiming ownership for a specified number of years) are not represented by a document. Such unrecorded transfers are always a challenge to the landman.

The landman or other title examiner can save a lot of time by obtaining an abstract of title prepared by an abstract company. An abstract of title is a book or file that contains copies or summaries of every recorded document affecting title to a particular tract. However, since the abstractor will not interpret the documents, the title examiner must know what the documents mean.

Because title is normally conveyed in writing, it is important to have a method of describing land area, so that the land conveyed by title can be determined on the ground. There are two basic methods of describing land areas:

1. The rectangular survey
2. The metes and bounds description

The *rectangular survey* is the simplest method to deal with. It was adopted by the United States Government shortly after the Revolutionary War and applied to almost all lands owned by the federal government west of the Appalachians. It is thus the system used in most of the oil-producing states in the Rocky Mountain-Prairie area except for Texas. Texas entered the Union as a republic and was never a federal territory, therefore the federal system does not apply there.

In the rectangular survey the surveyor lays out a north-south principal meridian and an east-west baseline to form a cross. He then measures off *townships* of 36 square miles each, 6 miles on a side (see Figure 1-1). The east-west measurement is then counted in units called *ranges,* and the north-south measurement is counted in units called *townships*. Thus, if you start at the intersection of the meridian and the baseline and want to describe the location of a township that is 3 units north and 2 units to the east (Figure 1-1) it would be described as Township 3 North, Range 2 East or simply—T3N, R2E.

Each township is then divided into 36 sections of one square mile each (640 acres), which are numbered 1 to 36 in the manner shown in Figure 1-2. A section can then be described simply by referring to a section number. The sections are not subdivided in the original survey but are easily subdivided by the owner into halves, quarters, etc. (See Figure 1-3.) Thus, the shaded area in Figure 1-3 con-

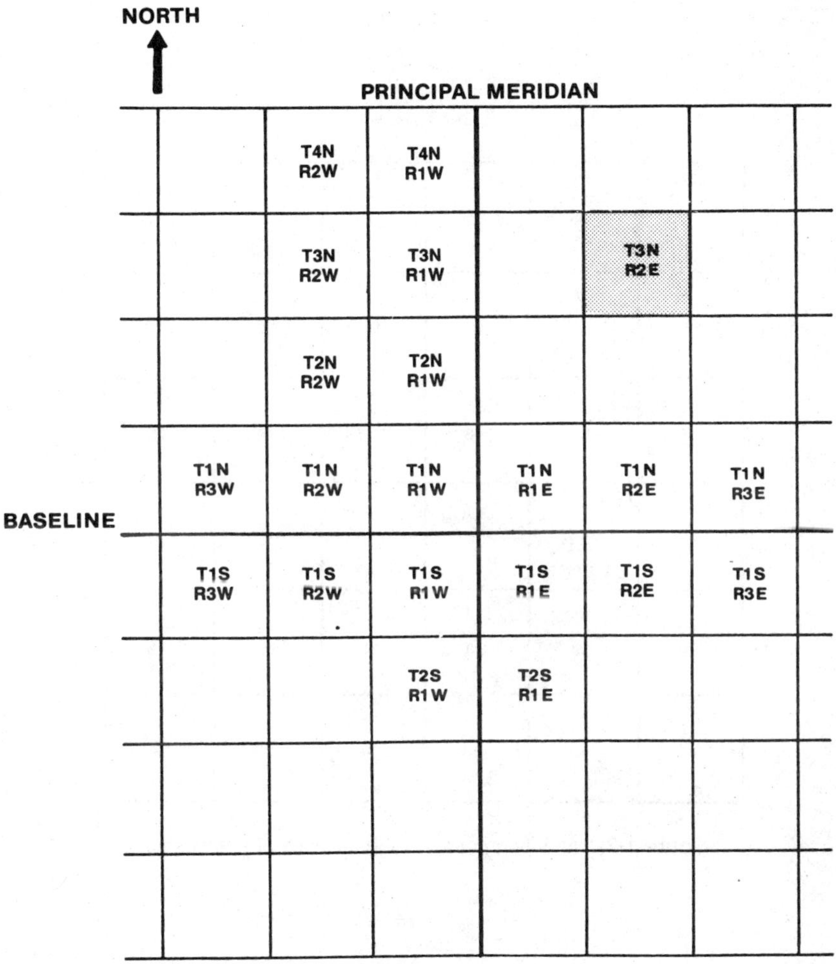

Figure 1-1. Example of a rectangular survey.

6	5	4	3	2	1
7	8	9	10	11	12
18	17	16	15	14	13
19	20	21	22	23	24
30	29	28	27	26	25
31	32	33	34	35	36

Figure 1-2. Numbering order of the sections of a township.

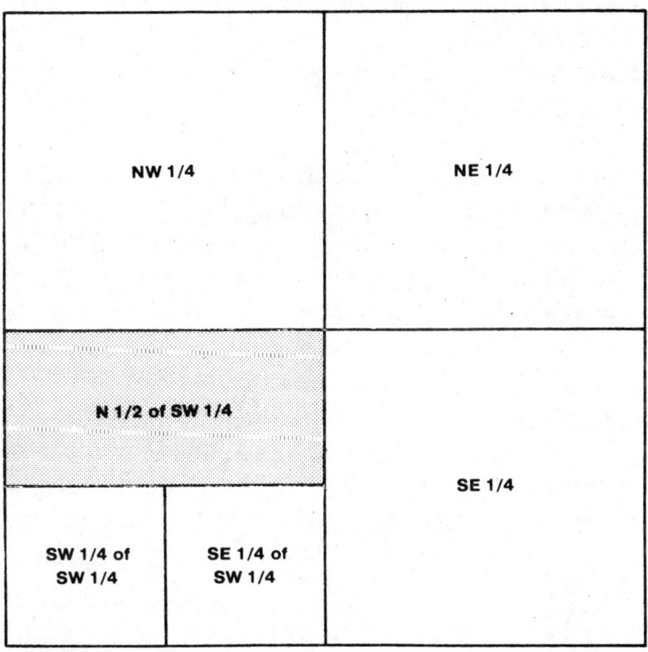

Figure 1-3. Example of ways a section can be subdivided.

sisting of 80 acres can be described as the north half of the southwest quarter, section eight, Township 3 North, Range 2 East. This is written N 1/2 of SW 1/4, Section 8, T3N, R2E.

A *metes and bounds description* is simply one laid out on the ground using:

1. Physical monuments (e.g., a river, a tree, a rock, a concrete marker)
2. Directions (e.g., North 30° East)
3. Distances

This is a difficult system to deal with because many early surveys were inaccurate to start with and the monuments have long since disappeared. This is the system used in most of Texas, although Texas has its own form of rectangular survey in the more recently settled areas. To make matters more confusing, in Texas everything is not traced to one federal government but to four different sovereigns, each with different methods of granting and describing land. Many early grants cover tens of thousands of acres in poorly described metes and bounds surveys, which sometimes use monuments such as trees that no longer exist.

2
The AFE

This and the following two chapters deal with the responsibilities of the petroleum engineer before the well is drilled. The rest of the book deals with the responsibilities of the consultant.

AFE stands for Authorized Field Expenditures. After the engineer reviews the prospect it is his job to write up the AFE for approval of the funds to drill the well. The AFE will include everything needed to survey, stake, drill, and complete the well in question. This chapter will go through an AFE, break it down and explain each item.

An AFE shows the operator or investor what it will cost to drill the well. If the well is turn-key, then the AFE will be more accurate than if the well is drilled using a daily charge. The AFE will itemize intangibles and tangibles incurred during the operation.

This chapter deals with drilling a vertical and a horizontal well. A capital H, enclosed in parentheses, appears before each item that involves horizontal drilling, not vertical. If both kinds of wells use the same item or service, then no **(H)** is used. Thus you can distinguish between the two.

Intangible Items on the AFE

1. *Surveying:* It is very simple to obtain a price for surveying a location. Simply call any reputable surveying company, give them the location, and tell them you need a firm bid. They will call you back with a price.
2. *Site preparation and roads:* The building of locations has always been a misunderstood part of drilling a well. Prices vary sometimes up to 100% especially in south Louisiana where board roads are necessary to drill in the rice fields. Always obtain three bids from reputable contractors that have worked in and around the area where you are to drill.
3. *Settling of land damages:* Settling land damages has always been a sore spot for the operator. Some landowners are very courteous, some are not, and some are just downright sorry. Trouble usually occurs when landowners do not understand that the owner of the mineral rights has the right to drill for oil or gas on the landowner's land. For the past 15 or 20 years, most operators have paid enormous amounts of money to landowners for crop damages and general problems. Paying a landowner a large settlement was an attempt to create goodwill. All it really did was create mass chaos for those responsible for land settling. Now that the oilfield has dried up along with the checkbook, the approach must be different. The standard offer should always be for just exactly the cost of replacing the lost crops or trees that had to be taken out, but that is it. However, the settlement will require considerable negotiating skill, so a well-

rehearsed tried-and-proven ''game plan'' will pay off.

4. *Clean-up:* Cleaning up a location is not as simple and cheap as it sounds. Many times a ''simple'' clean-up job has resulted in a tremendous amount of work and cash outlay. Make sure that your consultant is experienced in cleaning up locations and is familiar with the area that is to be drilled. Find a local contractor if possible. I say ''if possible'' because most contractors will not bid on clean up since there are so many variables included. Therefore, always include extra in your AFE to cover the unpredicted costs of cleaning up the location.

5. *The drilling contractor:* To figure the contractor's fees, you must first research the area you are drilling. You must study drilling bit records to find out how many days it took to drill different wells at different depths in that area. Then you must compare those depths and days with the depth of the well to be drilled. It is a simple procedure but sometimes time-consuming. The drilling contractor's fees become more of a factor on a tight AFE. On a loose AFE they are not as critical. In 1982 the cost of a drilling rig hit an all time low, and in 1983 many drilling contractors went out of business. Some of them were charging low fees just to have enough money to pay the interest on their loans. Most rigs are financed and sometimes the banks will repossess the rig even during a drilling operation. For the operator this is the cheapest time in recent history to drill a well, so you should get three or four bids from local contractors serving the area you are drilling. However, keep in mind that price does not absolutely guarantee success. In some cases, it is better to stick with older, established drilling contractors. Make sure they fur-

nish proof of insurance and in some cases, depending on the depth of the well, they should furnish a certified financial statement so you will be assured the bank will not repossess the drilling rig while they are on your hole.

6. **(H)** *Directional driller:* In horizontal drilling, this is the most expensive part of the whole operation. When horizontal drilling started, prices were high but livable, but then too many operators started to drill horizontal wells and a quick shortage of tools occurred and the directional drilling companies could name their price. This price fluctuation should start to level out since new tools are being built as quickly as possible. A lot of small operators have cancelled plans to drill horizontal, because of the cost and unavailability of the tools. The cost to drill a horizontal well now makes some operators run the other way. In 1991, the average price for MWD tools and an operator exceeded $11,500 per day. (MWD means measurements while drilling.) It is a very expensive MWD tool that gets charged for if it gets lost downhole due to a whiplash or pipe separation. When one is figuring an AFE on the directional team, a lot has to be considered.

First, consider the length of the horizontal hole. Say that the horizontal hole is 2,500 ft. The directional drillers get a setup fee, usually equal to a day's pay. Then you have to figure two to three days to drill the curve. I have heard horror stories where the curve took one to two weeks and was drilled 180° off. A consultant would be wise to try to understand everything the directional driller is doing, to act as a look-out for problems. Just one problem in the curve and most promoted wells are in trouble.

Then figure 250 to 300 ft per day on your AFE. So a 2,500-ft well should take ten days to drill, three days to make the curve, and one day to set up the equipment. This fourteen-day figure does not include trouble or kicks, so add three days extra to be safe. This rule of thumb should cover most chalk wells. However, some rules must be followed to achieve success. Upcoming chapters discuss safety features needed to ensure success. Horizontal drilling has many flaws, so a good consultant will be on his toes; but it is tough, considering everyone is trying to rip off the operator. In my opinion, the MWD system is best to use with directional drilling tools. It is a lot easier to use and safer than the steering tool system. Sometimes, however, the steering tool is all that is available. The disadvantage is the wait for each joint to be drilled before a survey can be made; however, by stopping the drilling operation, a survey can be run at any interval. It is a bit time-consuming, but the disadvantages are few compared to the benefits. I prefer the MWD system on all my wells. The only problem is the cost—it does cost more. Also I have looked at large and small companies in directional drilling, and I prefer the smaller companies because they are flexible. The big companies are not. When personnel from large companies arrive on location, the clock starts and you have to be ready. Smaller companies have greater flexibility; if the rig goes down, they can go off the payroll until you are back on-line. Large companies charge a big standby rate. Also the companies start the clock at midnight; so if you hire them at 8 p.m., you'll still pay for the whole day. I've always made sure that it was after midnight when they went on the clock. A good consultant can arrange for this to happen, and it will save the operator money. I am sure that as more tools become available, things will become more reasonable. The more people in the business, the cheaper the cost to drill will become, or they will collapse the whole industry.

7. *Drilling bits:* Obtain drilling or bit records from the area in which you are drilling the well (or call any bit company that can develop a bit program in your area) and an estimated cost. Compare the program with bit records, and try to estimate as closely as possible the cost and number of bits that will be needed. Remember, all bit programs are run on a computer—and I have never seen a 100% correct bit program from a computer. Therefore, it would be safe to add one or two bits to the program after actually comparing the bit records in that area.

(**H**) On horizontal wells, a rock bit is required to make the curve because of the time and rotation of the downhole motor. To be safe, figure two bits to make the curve. However, the Smith F-2 will normally do the job with one bit. Beware of directional drilling companies that tell you that their bit subsidiary has the best bit for the job. Any good rock bit works as well. To verify this, study the bit records in your area.

Once the curve is made, go with a PDC (diamond) bit. The MWD system will go about five days without a new set of batteries, so without tripping the bit, you can make some hole. Several companies make a superior bit that can be used to drill three to four wells. The cost is about $15,000, but over four wells this is cheap by any standard. Also most of the vertical part of a horizontal well is like digging a fence post hole, so some astute consultants turn to a used-bit program to save money. If you can find a reputable used-bit company, the savings are substantial. I've saved as much as $35,000 per hole by using good used bits. Just figure 80% life instead of 100%. Of course, you have got to watch closely toward the end of the bit run. The driller should keep his eyes on the torque gauge.

8. *Camp:* Most drilling consultants stay on location in mobile homes. Simply call any mobile home company specializing in renting mobile homes for drilling locations, and they will give you a price for 30 to 60 days on location. If you remember to include in your estimate water for the trailer and utilities, the rig-up and rig-down charge, and the moving charge you will have a pretty accurate figure.

9. *Rig move:* Most drilling rigs can be moved and set up for around $35,000 (based on current figures). This price, of course, is subject to fluctuations.

10. *Rental tools:* With every hole you drill, the operator needs rental tools. The most common are stabilizers for the surface hole and stabilizers for the well itself. In southern Louisiana, superchokes and degassers are normally rented, and on deeper oil pressure wells more exotic mud cleaning systems are needed. Simply call each rental company, obtain an estimated cost of each item, and put this on the AFE.

11. *Fuel:* Most rigs today have a daily charge that includes fuel, but if not, try to find out how many gallons of diesel the rig you are going to use consumes daily. If that figure cannot be obtained, just use a rough figure of 750 gallons daily to be safe.

12. *Drilling mud and chemicals:* Drilling mud is one item that does not vary too much in cost. Call any reputable mud company to get a bid on a complete drilling program.

13. *Cementing:* Depending on the design of your well, whether you set intermediate or liners, the price must be considered. Call any reputable cementing company and give them the specifications of your well, for example: a 12¼-in. hole to a depth of 3,000 ft setting a 9⅝-in. 36-lb casing. With that information the cement company can figure the cement and excess cement needed and give you a firm bid price excluding transportation. Always add an additional $1,000 for transportation to and from the location. Give them the same data on the long string, and they will be able to give you a price in a few hours.

14. *Coring and analysis:* If the well is to be cored in, say, two zones, call any coring company and they will be able to give you an estimated cost to use their tools.

15. *Drill stem test:* Call any reputable DST company and give them the estimated depth and pressures that you are going to encounter, and they will give you an estimated cost of running a test.

16. *Electric logging:* Prices for electric logging can be obtained by calling any reputable firm involved in electric logging in that area. Ask for a firm bid price.

17. *Mud logging unit:* Since the mud logging unit comes on locations at a predetermined depth it is necessary that the consultant estimate the amount of days the logging unit will be on location by reviewing drilling reports from that area.

18. *Trucking and hauling:* These costs should hold no surprises unless you have trouble at the rig—then the trucking can go up quickly. Normally, the hauling of pipe materials needed to drill the well will run about $4,000. This, of course, depends on the depth and the amount of hauling. By adding $1,000 to the

cost for surface and intermediate and long string you should obtain a reasonably accurate estimate.

19. *Casers:* Call a reputable casing company to run your casing. Because casing is so important, it is better to hire the older, established firms. The running of casing is one of the most important parts of drilling an oil well. The handling of casing by an old reputable firm should ensure a successful job.

20. *Float equipment, centralizers, liner hangers, etc.:* You can get this equipment as well as a price and an estimate for installing the equipment in the well by calling the companies handling the equipment in that area.

21. *Laydown machine:* This is used to pick up and lay down pipe in and out of the derrick. It is necessary after drilling a well whether you make a well or not. Many people fail to include this item in an AFE. Call any reputable laydown machine company for a bid on the depth you will be laying down drill pipe and drill collars. In cases where the operator pays for pipe inspection, the laydown machine will make it easier to handle the pipe and thus save you trouble on the pins and boxes.

22. *Squeeze cementing:* Squeeze cementing is necessary in just about every case in southern Louisiana. It is very hard to get a shoe to test in the surface hole. Most operators set surface pipe and work as hard as they can to drill it out and test it, but they do not leave enough time for the cement to set. Therefore, additional cement is needed at the seat to assure that the drilling operation will be successful if high pressure is encountered. So when drilling in southern Louisiana, always figure in one squeeze cementing job.

23. *Well-site supervision:* Well-site supervision is easy to figure. First, figure five days for location building. Add the estimated days of the drilling operation. Then figure eight days for location clean-up, depending on the area. Some locations can be cleaned up in one or two days, but some take a week. Obtain all the data possible in that area from local contractors.

24. *Insurance:* The operator needs to check how much insurance is going to cost on the well and then add that to the AFE.

25. *Geologist expenses:* Most operators, if they are drilling through a transition zone, hire a reputable paleontologist to locate the "bug." Smaller operators sometimes hire a well-site geologist. So, estimate the approximate days the geologist will be needed.

26. **(H)** *PWD (produce-while-drilling) equipment:* This is the new equipment that makes a horizontal well work. It requires a gas buster, a flare line, a tank system for the flowing oil, and the ability to separate the oil and water, to send the oil to the frac tanks for selling, and to send the water back to the mud tanks for repumping down the hole. It all sounds simple and it really is, but you can make it as exotic as you want. Most exotic systems are put together by inexperienced operators who are afraid the well is going to blow up. Exotic systems require more training, and by keeping the PWD system simple, it is easy to train rig hands and ensure success.

    Several safety rules should be followed in the use of PWD equipment. The main rule is that the equipment should be sturdy and tied down in case of large kicks. (See Chapter 29, "Blowout Control and Calculations," for a complete discussion of a kick.) The rig-up requires about five to six days before the operator begins horizontal drilling. A rotating head and a safety choke are necessary for the system to work; some operators, however, use two annular preventers instead of one. However, the rotating head and safety choke are used most. Choose a rotating head with a safety rating of 500 to 800 psi. Most PWD equipment can be figured at $2,000 per day on your AFE. Six frac tanks should be on location. If the area has a history of big kicks, more tanks should be available. A good engineer always knows where plenty of frac tanks can be found during horizontal drilling. Some companies now furnish all PWD equipment in package deals.

27. **(H)** *Rented drill string:* All horizontal wells require a string of 3½-in. pipe. Make sure it has been water-blasted and tumbled to get the slag out of the string. Also make sure that an inspection report is available. If it is not, get a good inspection firm to inspect the drill string before it leaves the rental yard. One small piece of slag can mess up an MWD tool, and to see the effect, just calculate what it costs to pull a string for a piece of slag. A lot of bad pipe is in the field, so beware of too cheap a price. You get what you pay for in rental pipe. Along with the pipe, you will need all the handling tools and drill collars as well as heavy wate drill pipe. Make sure that the MWD tool will fit the inner diameter (ID) of the pipe. Some drill collars have an ID that will not allow the MWD tool to go in and out of the string during the changing of batteries, so check with the MWD personnel about the ID needed before renting.

28. **(H)** *Trucking costs for transportation of oil off location:* This charge depends on how many kicks you have and how much oil is produced. Just figure in four kicks to be safe. Assuming that after a kick is

encountered, drilling continues, figure on 2,000 bbl per kick. If drilling does not continue, figure 6,000 bbl. This, of course, could exceed 20,000 bbl in some fractures. Just divide the estimated barrels by the transport load capacity.

29. **(H)** *The oil-treating chemicals:* After a kick has been encountered, the oil treaters will have to treat the oil before it can be sold. A barrel of treatment chemicals averages about $750 in 1991 prices. It is necessary to have the oil treated for market. Figure $3,000 to $4,000 per hole.

30. **(H)** *Guard services:* A gate guard is needed when the operator begins horizontal drilling. This is to keep all unnecessary people off location while kicks occur. It is simply too dangerous to pay less than 100% attention while kicks are going on. Gate guards cost from $150 to $300 per day.

31. **(H)** *Communication:* In vertical wells, good communication with the consultant, rig floor, toolpusher, and mud hopper is normally all that is needed.

    In horizontal wells, communication is very important all the time. A station needs to be set up with the consultant, toolpusher, rig floor, mud hopper, PWD tanks, directional driller's office, accumulator remote station, and guard gate. Wire units are more trouble than they are worth, so I started using CB radios with a 12-V adaptor; they work well, and you just monitor a channel that has no activity. A whole location can be rigged up for $1,200, and then the next well is free. A consultant needs a hand-held unit to carry at all times. A cellular phone with a fax unit is also needed. In some areas the cellular phone is not available yet, so a microwave unit is necessary.

32. *BOP testing:* For all wells, a blowout preventer (BOP) testing company needs to be hired to make

sure everything works. Horizontal wells are more important because you know the well will go through kicks. But any good engineer will require testing on all wells. The testing should be discussed with the drilling contractor before the well is drilled. If the BOPs will not test, then the operator should not pay for any downtime, while more equipment is ordered. Drilling contractors all have great BOPs until it's time to test; then the downtime starts, and usually the operator pays the bill anyway. So make sure that both parties agree when it comes to testing: All equipment will be tested, or the operator will not pay for downtime.

33. *Nipple up crews:* This is optional, and some engineers use this to save time. However, the nipple up usually goes more slowly than you want, so it is a tossup. I personally prefer a nipple up crew to work for my rig.

Tangible Items on the AFE

1. *Conductor casing:* Normally the conductor casing in southern Louisiana is driven through the ground with a drive hammer. Simply call a drive hammer company and give them the area. They will be able to tell you how much pipe can be driven and the estimated price.
2. *Surface casing:* Simply call for a bid from any tubular company.
3. *Intermediate casing:* Do the same as with the surface casing.
4. *Liner casing:* Do the same as with the surface casing.
5. *Well head:* This AFE is for a dry hole, not a producing well. (This manual does not deal with well completions.) So the well head may be purchased or rented. Of course, if a well is made, the well head must be

purchased. But in this instance just show a rental price rather than a purchase price. Also, add $250 to the rental price for reconditioning. Make sure that in the rental tool section of the AFE a wear ring is included—otherwise the well head could be permanently damaged by the kelly.

6. **(H)** *Storage tanks and separator and heat treating unit:* Many contractors can give you firm bids for the labor and equipment. Many operators build this equipment while the horizontal well is being drilled. Usually after a good kick, if it looks good, the operator can go ahead and assume he has a well. Not much chance is being taken after a good kick, since a good kick indicates a good oil reserve.

On all dry hole AFE's the cost of plugging and abandoning must be considered. Figure two days rig cost to plug and abandon most wells. Call for a bid on cement services and base your cement on the amount of productive zones that may be encountered. For example, if the geology shows that you will go through four sands, then figure four plugs. The sands may all be wet but it is necessary in most states to plug all four to isolate each zone from the others. So you figure in four bottom plugs, one plug 100 ft out and one 100 ft in at the surface casing plus a 10-sack plug at the surface. Also include the price of a welder to cut the well head and weld a metal plate on top of the surface pipe. All this can be estimated fairly easily and added into the AFE.

On most AFE's you can total up the tangibles and the intangibles and the P & A (plug-and-abandon) cost estimate and obtain a dry hole cost. As you can see, the oil field AFE is a valuable tool for estimating the cost of the well. If each category is broken down and careful research is done then the investor and the operators involved will be happy with the work.

Most AFE's with written bids take four to five days to complete. Of course, if you are in a hurry, they can be done in four to six hours. But if you want your AFE to be neat and complete, it takes four to six days to obtain all the data and properly submitted written bids.

Also in the AFE, the drilling contractors will be appointed. The following chapter will explain basically what the drilling contracts are.

3
Drilling Contracts

The rights and obligations of the operator and drilling contractor are set out in a detailed *drilling contract*. If possible, the consultant should be familiar with the contract for the drilling project. The contract sets out the items that the consultant, as a representative of the operator, is responsible for and the items the drilling contractor is reponsible for.

There are two basic types of drilling contracts. They are the *footage contract* and the *daywork contract*. In the footage contract the drilling contractor agrees to drill a hole of a certain depth at an agreed-on price per foot. The contractor assumes the risk of excess costs if the drilling falls behind schedule. In the daywork contract the drilling rig and crew are hired for a fixed rate per day or hour. This is no guaranteed rate of progress. In this type of contract much more risk and responsibility is shifted to the operator and thus to his consultant. The contractor agrees to perform in a workmanlike manner, but there is no guarantee of any amount of progress. Most operators prefer a footage contract because they can better estimate their costs. Then they hire a consultant to keep an eye on the operation so that the well will be drilled properly. However, many drilling contractors will not accept a footage contract in high-risk areas such as the Gulf Coast or offshore.

Most drilling contracts are signed on a standard form provided the International Association of Drilling Contractors (IADC). the IADC has three contract forms:

1. Daywork contract
2. Footage contract
3. International daywork contract

This book will examine the IADC Footage Contract (see Appendix A). Since the contract includes provisions for daywork payment when the rig is used for somthing other than drilling, i.e., coring, testing, etc., daywork agreements will also be studied.

If you are a consultant, or company man, you should be familiar with the IADC contract forms. Once familiar with the form you can quickly learn the terms applicable to your job by examining how the blanks have been completed and by noting any additions and deletions. Although all terms of the contract are important, some portions are of concern primarily to the home office. Those portions of the IADC Footage Contract that are of primary concern to the consultant will be examined here.

You will see as the contract is reviewed that there are many areas where the consultant can save the operator thousands of dollars without cutting quality, by simply avoiding waste and not paying for unneeded services. This is your job and can more than pay for your fee. Since you may be the only representative of the operator on the site, you may be asked to make decisions or grant waivers affecting legal rights. It is important for you to know when to notify the operator of developments and to ask for instructions. Your decision, even if beyond your granted authority, may be binding on the operator as far as outsiders are concerned and could create problems for him. The operator will highly appreciate the consultant who protects his rights, spots potential trouble, and keeps him informed.

Item 1—Location of Well: The importance of this item is obvious. Be certain that the contract relates to the well being drilled and that the location is correct. You are not expected to be a surveyor, but be alert to obvious discrepancies between the contract description and the rig site. Use common sense.

Item 2—Commencement Date: This can be very important to the home office. The lease may expire or other important legal rights may be lost if drilling operations do not commence by a given date. The consultant cannot run the operation until the hammering starts on location, but he can watch all delays closely, and he should never consent to any extension of time without consulting the operator.

Item 3—Depth: Notice in section 3.1 that there are three possible depths listed. The first is the agreed footage; the second is the depth to a given geologic formation which is the target (this may be either the zone from which production is expected or some landmark below which no zones worth testing are expected); the third depth gives the operator the right to stop at any time and set casing (this may be done if the operator finds a zone he is satisfied with or if the problems of further drilling become too great). Section 3.2 provides that drilling below the agreed footage is on a daywork basis. The contractor agrees to drill below the agreed footage but on a daywork basis, and the guaranteed price per foot does not apply. The contractor need not go below the maximum depth set forth in section 3.4. The contractor will have sent a rig suitable for the planned depth, but it may be inadequate for greater depths, and the contractor will not assume responsibility for drilling deeper. The agreed footage is always the consultant's target. He should never set casing short of this or drill beyond it without instructions from the operator.

Item 4—Work Stoppage Rate: Note in section 4.3 (d) the obligation of the operator to provide road access suitable for ordinary road vehicles. As the on-site consultant you must see that this is done. As provided in section 4.3 (b) and (c) if the rig is available and cannot be moved on or off the site because of inadequate roads the operator must pay the work stoppage rate for as long as it sits there. This is money thrown away.

Item 4.4—Repairs: If the rig is drilling on a footage rate the contractor is responsible for repairs, but if it is on daywork the operator must pay daywork rates for the rig while it is being repaired (but not the costs of repairs). Each contract specifies a minimum amount of downtime that will be allowed with no dock in pay to the contractor. After that time, the operator docks the contractor for each hour the rig is down.

Item 4.5—Standby Rate: If the rig is standing idle the contractor is paid at this rate. Even if drilling is on a footage basis, if the rig is idle because, for example, the casing did not arrive, it costs the operator money. The consultant must coordinate all services and supplies on site when needed, because idle rig time is very expensive and is wasted money.

Item 4.6—Reimbursable Costs: The drilling contractor will normally be very happy to provide any service that is supposed to be provided by the operator. He will simply purchase the goods or services elsewhere and be reimbursed for his total cost plus the agreed percentage markup, typically about 10%. Do not let him. You can throw away thousands of dollars in extra costs for items you are being paid to handle directly.

Item 4.7—Daywork: It is important to know that even when drilling on a footage basis the contract reverts to daywork rates when certain services are performed. Read section 4.7 (c) to see what will cost extra. Note in section 4.7 (b) that if the casing or cementing fails, all restoration work is paid for by the operator at daywork rates over and above the footage rates. You, not the drilling contractor, must ensure that you have good casing and cement. Section 4.7 (c) provides that while performing all extra services (running casing and cementing are not extra), the rig is paid for at daywork rates. Even though the operator pays the cost of casing and cementing (see sections 6.16 and 6.26 of Exhibit A) he does not have to pay extra for rig time during these operations. For testing, etc., he pays for the test as well as for the rig time used.

Item 12—Difficult Formations: This section contains a list of difficult formations that may be encountered. You should read it carefully. In each case the risk and expense are shifted back to the operator by daywork rates. The consultant must stay extra alert during any of these conditions and should always seek special instructions from the operator. In certain cases, especially when granite or other igneous rock is struck, the operator may decide to drill no further and either to set casing or to abandon.

Item 15—Sound Location: As we mentioned earlier, you must see to the preparation of the location for the rig and to the placement of the conductor pipe.

Item 18—Lost Equipment: The contractor is responsible for loss to his equipment above ground. While drilling on a footage basis the contractor is also responsible for loss of in-hole equipment, but when the contract shifts to daywork rates the operator is responsible for loss of in-hole equipment. Since the contract shifts to daywork when problems are encountered, the operator becomes responsible for loss at the time of highest risk. The same applies to loss or damage to the hole itself. While drilling on a footage basis the contractor takes the loss if the hole is lost or damaged; on a daywork basis the operator stands all losses to the hole. This only applies to loss of the hole itself, however. The operator is always responsible for any damage to any formation or loss of oil or gas (see section 18.8).

Exhibit A: Perhaps the most important part of the contract for the consultant to be familiar with on each job is Exhibit A. As you can see this specifies in detail what will be provided by each party. You will find the casing program (Item 1), the mud program (Item 2), the straight hole specifications, ID, the maximum permitted deviation from straight vertical and the required surveys (Item 3), the equipment furnished by the contractor (Item 5), the equipment furnished by the operator (Item 6) and certain items to be furnished by either party as may be designated (Item 7). Exhibit A is a working document for the consultant, and he can hardly do his job without knowing what it provides.

One final word, the IADC contract was prepared for drilling contractors and tends to protect the drilling contractor and not the operator. There is nothing wrong with this, and in fact this is true of most printed form contracts. It is not, however, like the law of the Medes and the Persians which could never change. The parties are free to make any additions, changes, or deletions they agree upon. Always look for strikeovers or additions when reviewing a contract, as well as checking how the blanks were completed.

One of the largest problems I have encountered is the blowout equipment not testing. It is best to come up with a plan before the well is drilled. Too many operators have been misinformed about the condition of the BOPs, chokes, and annular preventers. If they do not pass a pressure test, the rig needs to be off the payroll until they do.

Another major problem is that all drill collars must be inspected before the well is drilled. This needs to be specified at the prespud meeting. If the rig is new to your company, then even the drill pipe needs a quick check. It is easy to do—hire an inspector to spot-check the pipe before the rig arrives at your location. Most contractors try to use the pipe without a proper inspection, and the operator always pays the price of downtime and fishing jobs. Even though I have a lot of friends in the drilling contracting business, I still never take anybody's word about pipe. Always pay for a spot check. It will save you money.

4
The Drilling Prognosis

Oilfield work is rich in its own terminology. Understanding the language is a major part of the battle of learning petroleum technology. It is therefore important that the student study the glossary and become familiar with the terms peculiar to the oilfield.

By using examples and pictures, this section will explain how to drill a well. The example well will be drilled to a total depth of 9,200 ft. Remember, these are just examples, and the formulas given in the text are for all wells at all depths and all diameters.

Most wells in the world are drilled by drilling consultants because most oil companies do not have a staff trained to handle the job themselves. The oil company, or simply, the operator, is the person or company who has the rights to drill for and produce petroleum products on a particular site. The operator may be a major company or an independent. Most drilling in the United States is done by small independents. The company will normally seek a number of investors to finance the drilling. Once financing is secured, a drilling consultant is hired to take over. The consultant works from a drilling prognosis prepared by the company engineer. The drilling prognosis is simply a plan by which to drill the well.

The drilling prognosis is given to the consultant prior to his arriving on location, and will contain the following items:

1. Lease and well name
2. Directions to the location
3. Company involved as operator (consultant's boss)
4. Drilling contractor to be used
5. Mud service company to be used and the mud program
6. Services to be used (called vendor's list)
7. Mud loggers to be used
8. The drilling procedures and practices
9. Company officials to be called in case of emergency
10. Bit records in location area
11. The drilling contractor's contract with the company, including an inventory of items furnished by contractor and operator
12. Drilling permit issued by state or federal authorities
13. The location layout (Figures 4-1 and 4-2)
14. **(H)** The directional and MWD companies to be used
15. **(H)** The PWD equipment to be used
16. **(H)** The frac tanks to rent for location

Items 1 through 4 are self-explanatory.

Item 5: The mud program will list mud reports of other wells that have been drilled around the location. It will tell you what has been successful and will furnish a list of weights and viscosities to use at certain depths. It will also detail what chemicals need to be added at each depth.

Example

Depth (ft)	Weight (lb/gal)	Viscosity
0–3,000	8.9	40–45
3,000–5,000	9.0	40
5,000–7,500	9.3	38–40
7,500–9,200	9.5	38–40

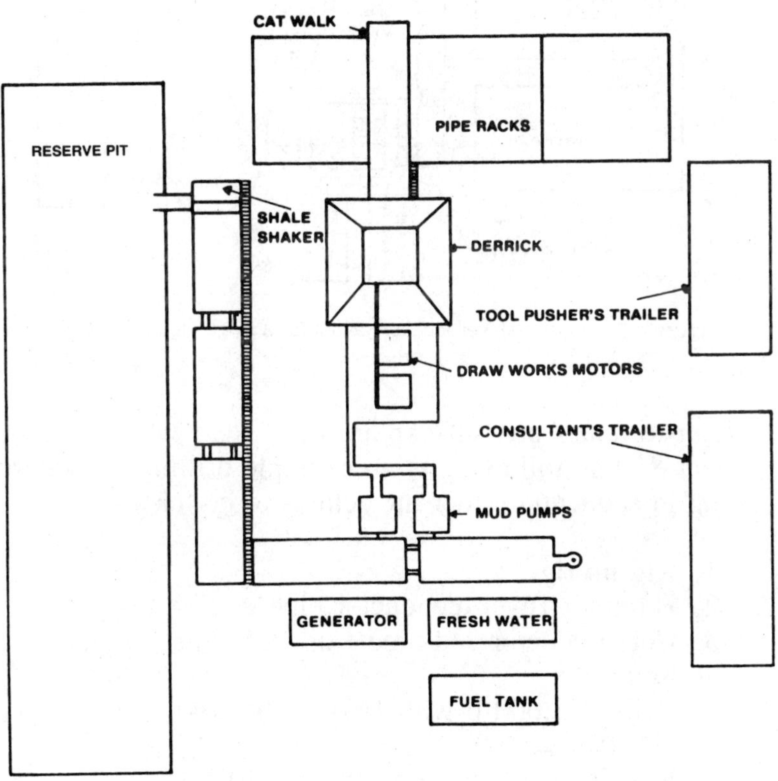

Figure 4-1. Vertical well location layout.

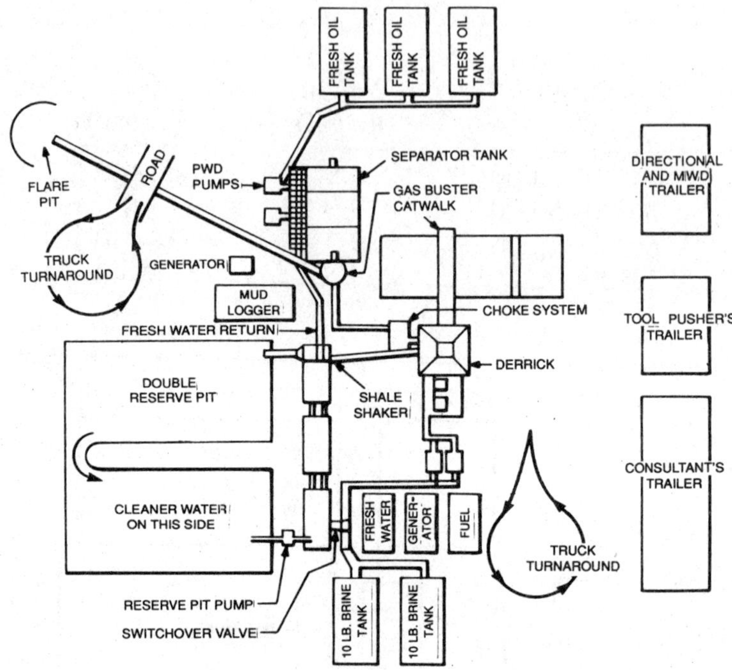

Figure 4-2. Horizontal well location layout for a single separator tank.

Items 6 and 7 are self-explanatory.

Item 8: The following is an example drilling procedure listing in sequential order the actions to be taken.

1. Rig up rig.
2. Hammer 16-in. conductor pipe to 120 hits per ft.
3. Cut conductor and nipple up flow lines to pits.
4. Spud in.
5. Drill ± 3,000 ft with 14¾-in. bit, run survey every 500 ft.
6. Run ± 3,000 ft with 10¾-in. 40.5 lb per ft K-55 casing with guide shoe, float collar, six centralizers, and one cement basket. Install three centralizers

every other collar. Install cement basket in conductor with two centralizers above on collar or lock ring.

7. Cement with 2,200 sacks (sk) Class H cement with 2% CaCl.
8. WOC (wait on cement).
9. If no cement returns, one-inch the hole.
10. Cut casing. Weld on 10-in. 1500 series casing head.
11. Test casing head to 1,000 psi for 15 minutes.
12. Nipple up BOPs, annular preventer, choke manifold, and superchoke.
13. Test BOPs 5,000 psi, annular preventer 2,500 psi, choke manifold 5,000 psi.
14. Make up the bottom hole assembly, and 9⅞-in. bit without stabilizers.
15. Trip in hole and tag cement, test 1,500 psi.
16. Drill out float collar, test casing, 1,500 psi.
17. Drill out guide shoe and 10-ft formation, test to 13.5 EMW (equivalent mud weight).
18. Drill with 9⅞-in. bit until it wears out. Survey every 500 ft.
19. Trip out of hole and pick up stabilizers.
20. Drill ± 9,200 ft.
21. One degree deviation per 1,000 ft not to exceed 7°.
22. Condition hole and run logs, ISF-CNL-FDC-Dipmeter, and RFT tool, and core guns 40 shots.
23. Plug and abandon or run pipe.
24. Run ± 9,200 feet with 7-in. 23 lb/ft K-55 with guide shoe, float collar and 10 centralizers.
25. Cement 280 sks Class H with 2% CaCl.
26. Cut casing, nipple down, and release rig.

The horizontal well prognosis:

1. Drill rat hole, mouse hole, and start hole.
2. Rig up rig so flare line is downwind. Choose the prevailing wind, and set up location so the flare line is in a safe location.

 3. Pick up a used 12¼-in. bit, spud in, and drill to ±
    750 ft. Run survey every 250 ft.
 4. Run ± 750 ft 10¾-in. 40.5 lb per ft K-55 used casing
    (if used can be found) with guide shoe, insert, eight
    centralizers, and one cement basket. Run six cen-
    tralizers every other collar, starting at the first collar
    on the bottom; then last three joints; then put one
    centralizer below the cement basket and one above.
    Lock the centralizer with lock ring.
 5. Cement with Class H cement with 2% CaCl. Figure
    100% excess.
 6. If no barrels of cement are circulated to the surface,
    then make the hole 1 in. Get together with cement
    engineer on location to assess requirements.
 7. WOC (wait on cement).
 8. Cut casing. Weld on 10-in. 1500-series casing head.
 9. Test casing head to 1,000 psi for 15 minutes.
10. Nipple up BOPs, annular preventer, choke manifold,
    and superchoke.
11. Test BOPs to 5,000 psi, annular preventers to 2,500
    psi, and choke manifold to 5,000 psi.
12. Make up the bottom hole assembly and 8¾-in. bit
    without stabilizers.
13. Trip in hole (TIH) and tag cement; test casing to
    1,000 lb.
14. Drill out insert, and test casing to 1,000 lb.
15. Drill out guide shoe and 10-ft formation. Test to 10
    lb EMW.
16. Drill with 8¾-in. bit to top of Carrizo sand. Then
    slow down bit to 70 to 90 rpm until the bit wears
    out. Survey every 500 ft. Watch very closely for
    torque in Carrizo sand.
17. Trip out of the hole (TOH) and pick up stabilizers
    and new bit.
18. Drill to ± 6,700 ft.
19. Condition hole, short trip through Carrizo sand, and

circulate bottoms up. Build up viscosity to 60 to 90. Your mud engineer can give advice on this question.

20. Pull out of the hole (POH) to run logs.
21. Run electricity logs.
22. Pickup Monel collar and TIH to bottom, circulate bottoms up, short trip back through the Carrizo sand, then TIH to total depth (TD) and circulate bottoms up.
23. Make up multishot gyro, and drop in drill pipe.
24. TOH to run casing, lay down 4½-in. string drill pipe and collars, and recover gyro tool.
25. Make sure multishot worked before running casing.
26. Run 7-in. 23 lb/ft K-55 used casing with guide shoe, float collar, stage collar, two cement baskets, and ten centralizers. Run six centralizers every other joint at the bottom. Place stage collar (DV) tool per rules for protecting water sands. Place cement baskets 100 ft in and 100 ft below the top of the surface casing with centralizers, using lock rings.
27. Cement first stage with Class H cement with 2% CaCl, drop bomb and open second stage, circulate for about 4 hours, then cement second stage with light cement with returns to surface. Record amount of returns to surface on cement record.
28. WOC 12 hours.
29. Nipple down BOPs and cut casing.
30. Install B section.
31. Nipple up stack and add rotating head: add gate valve from rotating head to shale shaker. Must be 1,500-lb test.
32. Nipple up choke and lines to PWD equipment. Hook up gas buster and flow line to separator tanks. (Most of this should have been done during drilling of the vertical hole.)
33. Rig up frac tanks with 10-lb brine hooked up to mud pumps.

34. Rig up PWD tanks to accept oil.
35. Rig up frac tanks to accept clean oil for sale.
36. Test BOPs, annular preventer, choke system to same value as first test.
37. Train crews on PWD equipment, including rotating-head operation.
38. Pick up 3½-in. rental string and 6⅛-in. bit and TIH and tag the DV (stage) tool. Test casing to 1,500 lb.
39. Drill out the DV tool slowly, and test casing to 1,500 lb for 15 minutes.
40. Drill out float collar and tag the guide shoe. Test casing to 1,500 lb for 15 minutes.
41. Drill out guide shoe and 10-ft new formation. Test formation at shoe to 11.5 EMW.
42. TOH and change bit.
43. TIH and drill to ± 7,400 ft.
44. Condition hole and TOH for logs to determine chalk sections.
45. TIH open-ended to cement open hole.
46. Cement open hole and 100 ft in 7-in. casing and 200 ft out.
47. TOH with drill string.
48. WOC.
49. Rig up directional driller and MWD tools.
50. TIH and tag cement.
51. Drill to a depth determined by directional driller.
52. TOH to pick up directional tools.
53. TIH with directional tools.
54. Drill curve.
55. TOH to pick up horizontal tools.
56. TIH with horizontal tools and drill 2,500-ft horizontal.
57. Kill well to TOH.
58. TOH to 7-in. casing, check for flow. If well is dead, TOH slowly filling hole. If well is flowing, circulate

pipe with brine (weight can be determined by the consultant). Keep it as light as possible.

59. Run wireline packer 60 ft inside 7-in. casing and set.

60. Nipple down stack and clean mud tanks. Release rig.

Items 9 through 12 are self-explanatory.

Item 13: (**H**) The horizontal layout, as you can see, is quite a bit more complex. It looks like a small oil refinery. A consultant must check the prevailing winds to set the location so that the flare line is downwind in case of H2S gas. Also make sure all roads out are not in harm's way if the rig should blow up. All these factors need to be taken into consideration by a competent consultant or engineer. Setting up a horizontal location is serious business.

Items 14 through 16 are self-explanatory.

The bit program is designed to show you what bits have been successful in the area you are going to drill, the rate of penetration, the RPM and WOB (weight on bit), the hours on the bit, and the grade the bit came out of the hole.

The drilling prognosis is the consultant's bible on location. Every phase of the operation is covered in the prognosis. Refer to it often. The toolpusher, mud engineer, and cement engineer should have a copy with them to correlate with yours.

The cement program will tell you what combination of cement to use on location and where to place the centralizers.

The casing program will show what grade, weight, and thread design is to be used in the wells. Most deeper strings have several weights that need to be layed out before going in the hole.

Figures 4-3 and 4-4 show the rigs with casing in the ground. The deeper you drill, the smaller the hole becomes. The prognosis will tell when the pipe is to be run and at what depth.

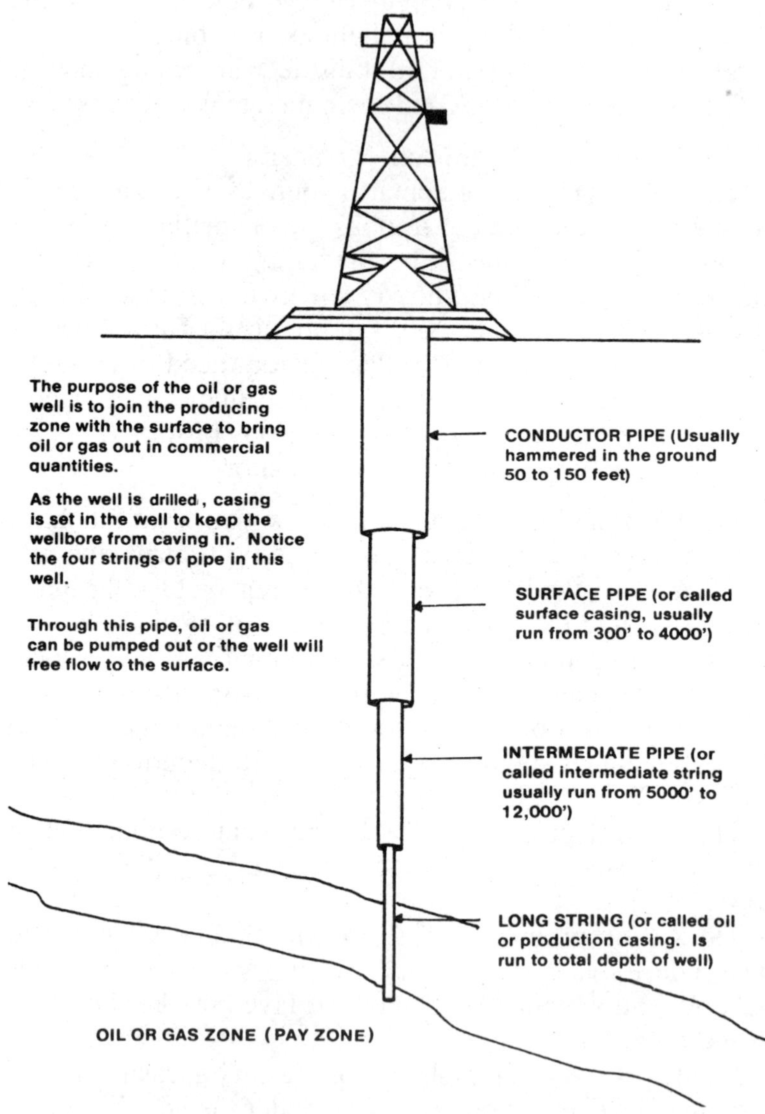

Figure 4-3. The piping needed to get oil or gas from a vertical well.

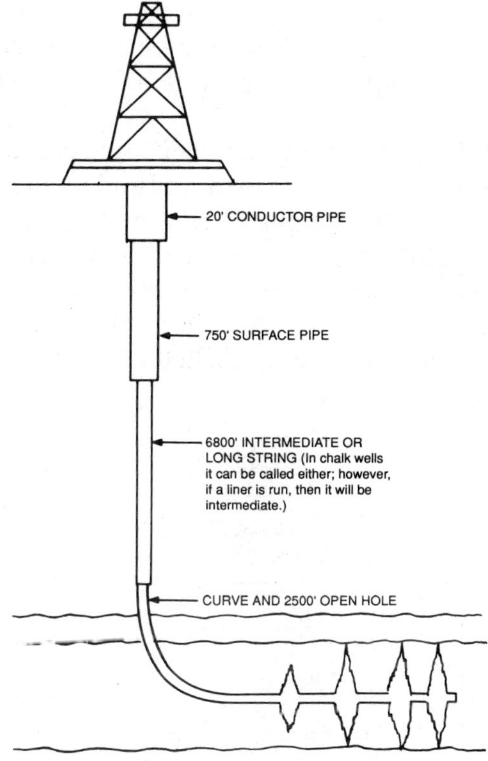

Figure 4-4. The piping needed to get oil or gas from a horizontal well.

What Is a Drilling Rig?

A drilling rig is a large and complex mechanism whose basic function is relatively simple. Its function is to rotate a string of drill pipe and drill a hole in the ground. It must also pull the drill pipe out of the hole for drill bit changes and run pipe back into the hole. The rig is really a giant crane for lifting and lowering drill pipe, with a rotary table to rotate the pipe. That is why it is called a *rotary rig*. A more ancient method known as the *cable tool rig* is not used on deep or high pressure wells and will rarely be used any more, except to hold a lease until a rotary rig becomes available.

The drilling rig must be able to perform certain secondary functions such as circulating drilling fluid to clean the wellbore and support the weight of the drill string so that the weight on the bit can be controlled. For example, if the string weighs 200,000 lb and only 30,000 lb of weight is needed on the bit, the rig must support 170,000 lb of the string while the well is being drilled (see Figure 4-5).

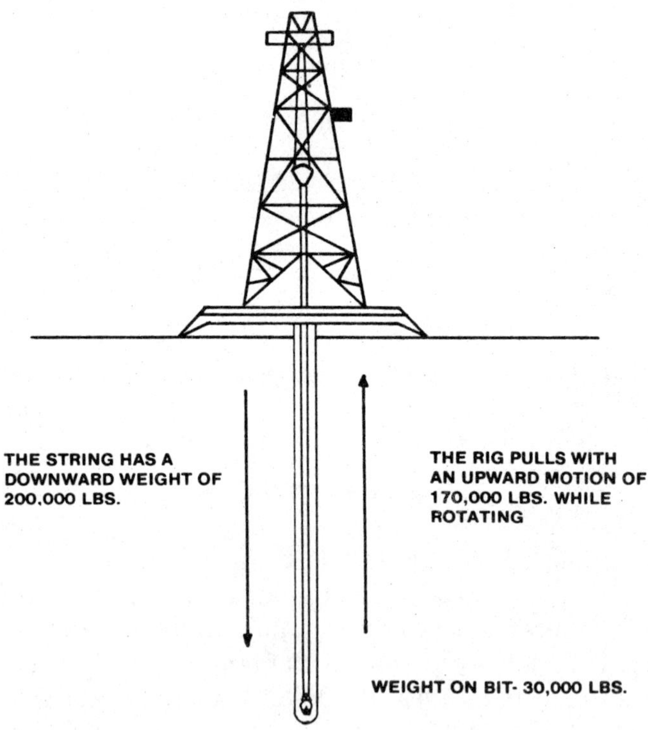

Figure 4-5. The rig supports the weight of the drill string.

5
Key Rental Items

When the consultant arrives on site, he will need to order out the rental tools or equipment to be used on the location. The following is a list of the most important items. It does not include many small things that come and go on location. Some needed items include:

1. Superchoke
2. Degasser
3. Shale shaker
4. Mud hopper
5. Intercoms
6. Mud loggers
7. Jars and stabilizers
8. Shock sub
9. Flow show, pit gain, and gas indicators
10. Mobile home
11. Mobile phone
12. Wear ring and puller

13. **(H)** PWD (produce-while-drilling) equipment
14. **(H)** Directional tools and MWD (measurements-while-drilling) tools
15. **(H)** Rotating head
16. **(H)** Rental string and handling tools
17. **(H)** Frac tank rentals

Superchoke—A hydraulic valve that is operated from the floor or from a station some distance from the rig. It is a part of the choke manifold system and uses a sand choke to control pressure where there is a chance of high-pressure gas. It is normally charged out with a 30-day minimum which makes it an expensive but necessary item. (See Figure 5-1.)

Degasser—Device that circulates mud from the mud tanks and separates the gas from the mud. It is very necessary when gas is expected. Most service companies that handle the item will solicit the consultant's business. (See Figure 5-2.)

Shale shaker—Device to separate the cuttings from the return mud. It should be rented if the rig does not have one on location. (See Figure 5-3.)

Mud hopper—A tank furnished by the mud company to move bulk additives to the mud system faster. If high-pressure gas is expected it is best to have one or more on location, because the mud hopper furnishes bulk barite faster than if sacks are used.

Intercoms—A much needed item for proper communication around the rig. Master units need to be placed in the following locations:

- Driller's control
- Companyman's shack
- Toolpusher's shack
- Mud engineer's shack

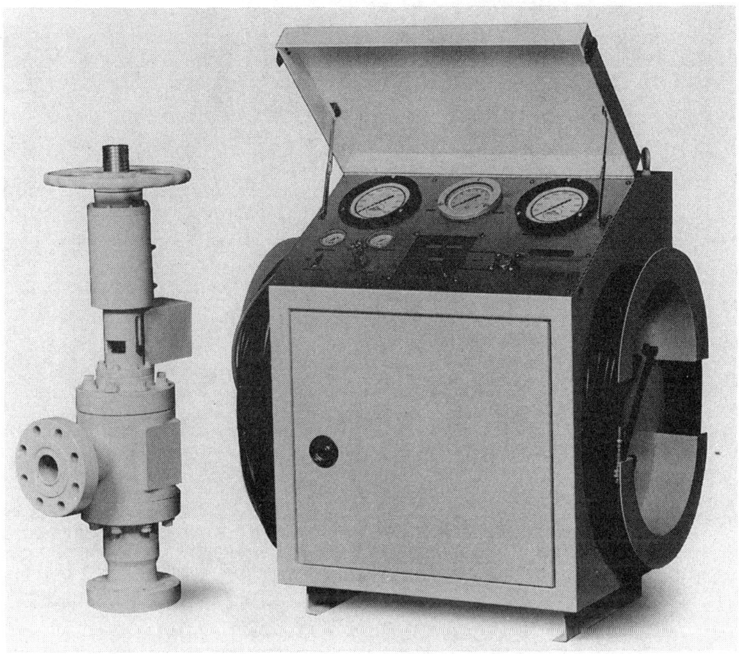

Figure 5-1. The superchoke system. (Courtesy of Sweco Oilfield Services, a division of Environmental Procedures Inc., Houston, TX)

- Derrickman's mud hopper
- Monkey board (on high pressure wells)
- Mud logger's shack
- **(H)** PWD separator tanks (Figure 5-4)
- **(H)** Directional and MWD trailer
- **(H)** Guard shack

Mud loggers—A trailer loaded with recording devices to monitor the gas and the geology. These devices are hired by the operator and are not the consultant's responsibility. Just ensure they are hooked up properly and their time on location is recorded for billing.

Figure 5-2. The degasser takes the gas out of the drilling fluid and returns the degassed mud to the mud tank. (Courtesy of Sweco Oilfield Services, a division of Environmental Procedures Inc., Houston, TX)

Drilling jars and stabilizers—These are not always used. The requirements for drilling jars will be listed in the prognosis. When required, make sure they are on location when the BHA (bottom hole assembly) is made up. Jars allow the pipe to be jarred out of the hole when the pipe becomes stuck. Stabilizer's hold the BHA straight in the hole. In my opinion a drilling jar should be used on all wells below 3,500 ft and in some areas that have hole problems. It is good insurance because if you get stuck, you may be able to free the pipe. I prefer the Dailey drilling jar (Figure 5-5). I have never had a problem with it, and if you are in

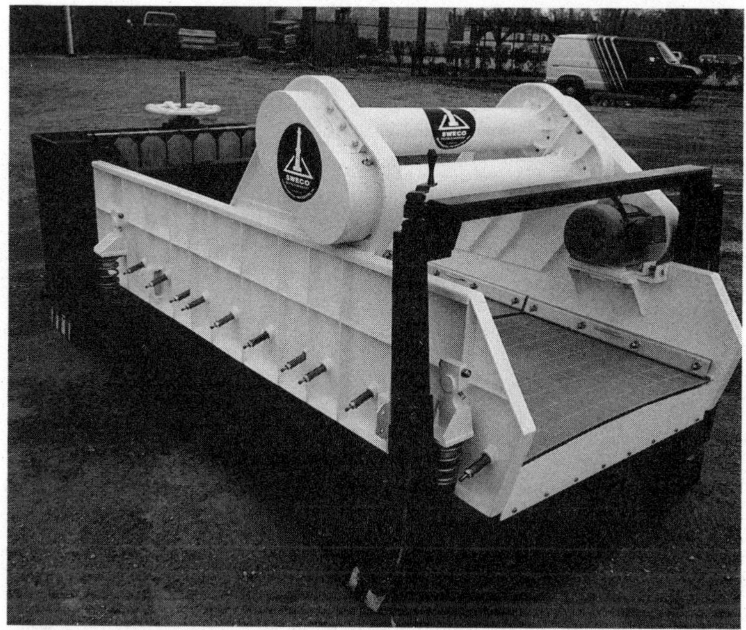

Figure 5-3. The shale shaker separates the cuttings from the return mud. (Courtesy of Sweco Oilfield Services, a division of Environmental Procedures Inc., Houston, TX)

a bind, the company will send someone out to help the consultant.

Shock sub—Should be used on every well below 7,000 ft to increase bit life and to eliminate bouncing of the drill stem in the hole. It serves the same purpose as a shock absorber on a car.

Flow show, pit gain, gas indicators—Devices to keep the operator informed of mud flow and pit gain or loss. The gas indicators register the units of gas in the mud. The service company will install the units and make sure they are functioning properly. These items need to be on every location where lost circulation and gas kicks may be encountered.

Figure 5-4. Crew member working the separation tank. (Courtesy of Davenport Horizontal Drilling Consultants, San Antonio, TX)

Figure 5-5. Drilling jars. (Courtesy of Dailey Petroleum Services Inc.)

Mobile home—Call a mobile home rental service. In most oilfield areas they are easy to obtain. The rental company will bring it out and set it up. Always make sure they bring a fresh water system. After the job is finished they will pick it up.

Mobile phone—Call any communication company and have them set up a cellular phone or microwave in the company trailer. Try to get at least three channels, so that if one channel is overloaded, you can switch to a different one.

Wear ring—A device to keep the kelly from wearing out the bradenhead while it is turning. A ring puller will also have to be rented with a wear ring. It can be installed with a joint of drill pipe and requires a puller for removal.

(H) *PWD equipment*—The produce-while-drilling (PWD) equipment covers all equipment necessary to drill horizontal. The equipment includes the gas buster, separation tanks (Figure 5-6), flare unit and pipe pumps to pump water

Figure 5-6. Separation (or skimmer) tanks. (Courtesy of Sweco Oilfield Services, a division of Environmental Procedures Inc., Houston, TX)

back to the mud tanks and oil to the frac tanks to sell, frac tanks to furnish brine water, frac tanks to sell oil, generators and mobile lighting to light the area at night, and all the lines to hook up the system. Since this is such a new technology, Sweco Oilfield Services, a division of Environmental Procedures Inc., has got a package deal that includes all the above. If you have to rent and set everything up separately, it will take a lot of time and often you will spend more per item. When the rig is ready to move, the company will move it for you to the next location.

(H) *Directional tools and MWD tools*—The directional drilling tools and MWD tools should be ordered by the engineer before the well is begun. Without these tools a well cannot become horizontal. The tools include downhole motors, stabilizers, subs, and surface directional devices. The MWD tools fit in position in the string; they can be retrieved by wireline to change the batteries. They are very expensive and fragile. In picking a company the engineer should look at its past performance and most important the reputation of the directional driller. This is very important when the curve is drilled.

(H) *The rotating head*—The rotating head (Figure 5-7) is a device that allows you to drill under pressure and to strip in and out of the hole. There are many rotating heads for rent, however; personally I like the rotating head made by Williams Tool Co. Also Williams has developed a dual rubber system that holds up to 900 psi and is working on a 1,500-plus unit. I am sure that soon horizontal wells will be drilled everywhere and in many different situations.

(H) *Rental string and handling tools*—To drill a horizontal well, a 3½-ft drill pipe string must be used, so a good string must be located. There are many rental companies to check out. Since supply of the pipe is limited, it must be lined up as soon as possible.

Williams Tool Co., Inc.

WILLIAMS DOUBLE RUBBER HIGH PRESSURE ROTATING CONTROL HEAD

Williams Rotating Control Heads are designed to operate to 500 psi nominal and 1,000 static pressure. With new drilling technology and horizontal drilling activity. Williams Tool Co. have added a second stationary high pressure rubber.

The stationary assembly carrier rubber diverts the high pressure and allows drilling in higher pressures (400 - 900 psi) a pressure gauge at the driller console monitors pressure between the two rubbers. When the gauge reads in excess of 400 - 500 psi. The bottom assembly carrier stripper rubber can be quickly changed and drilling continued.

Specify: Bottom Flange Size and Series when ordering.

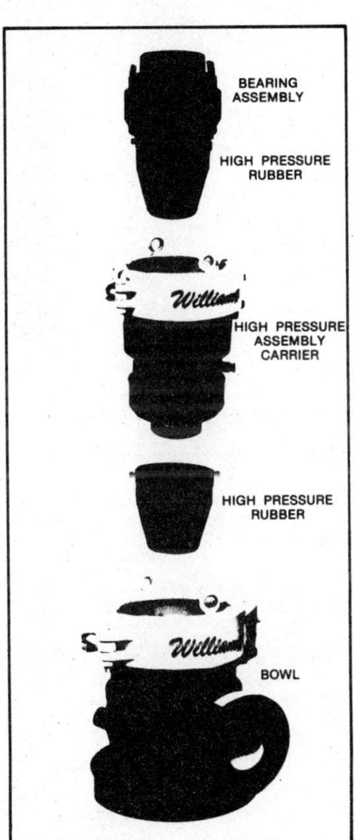

Figure 5-7. Rotating head. (Courtesy of Williams Tool Co., Inc.)

(H) *Frac tank rentals*—Frac tank rentals are very important to the success of a horizontal well. These days they are hard to find in certain boom areas, so the engineer or consultant needs to line them up in advance, if possible. Sometimes it is necessary to pay a little extra to guarantee the supply.

6
Rig-up, Spud-in, and Setting Surface Casing

If you are the consultant, the rig hands may already be rigging up by the time you arrive at the rig site. The size of the rig and the weather will determine how fast the spud-in operation will begin. On the Gulf Coast a drive hammer crew is hired to drive the conductor pipe. Normally the conductor is driven until it takes 100 to 130 hits per foot. The depth it is driven will depend on the diameter and thickness of the pipe. The pipe will usually drive from 50 to 160 ft depending on the geology of the area. The hammer crew should know to what depth they will drive in each area. Most hammer companies bring their own welder to weld the joints; if not, one will need to be hired. (See Figures 6-1 and 6-2.)

In hard rock areas a small spud unit drills a spud hole into the ground, where a temporary conductor is set, and a flow line is welded to the conductor and run to the mud tanks (see Figure 6-3). After the surface casing is set the temporary conductor is cut off and taken out of the cellar.

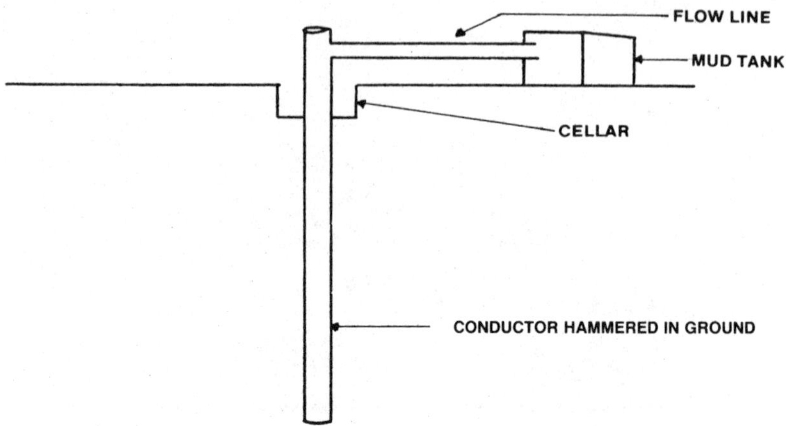

Figure 6-1. The conductor pipe hammered into the ground.

Figure 6-2. A drive hammer crew drives the conductor pipe into the ground. (Courtesy of Davenport Horizontal Drilling Consultants, San Antonio, TX)

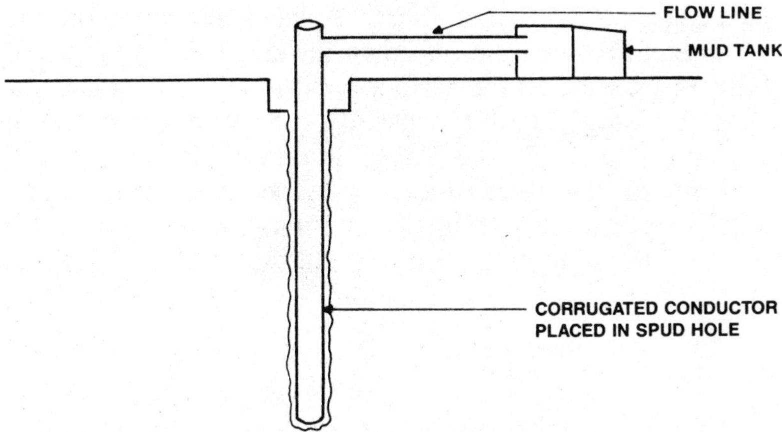

Figure 6-3. In hard rock areas a spud hole is drilled, and a temporary conductor pipe is set in the hole.

In horizontal drilling, the start hole is lined with culvert pipe just to guide the first drill collar and bit to spud-in. No flow line is needed because of the small depth involved (see Figure 6-4).

After the conductor is welded up, the crew mixes water and gel and then spuds in. The main elements of the drill stem will be the bit, drill collars, stabilizers, and drill pipe

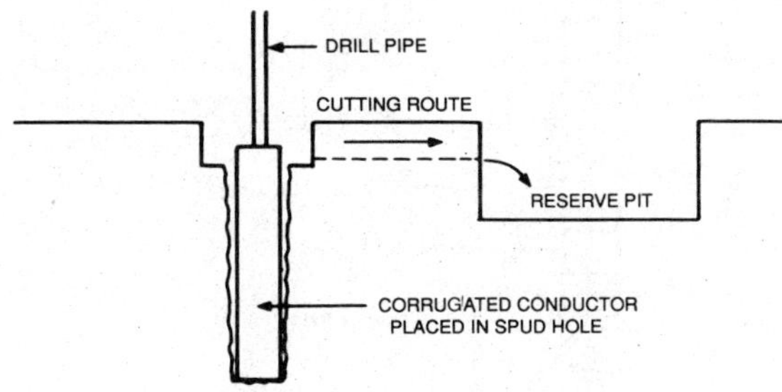

Figure 6-4. Horizontal start hole.

(see Figure 6-5). Drill collars are placed above the bit and add weight to send the bit down faster. Normally two stabilizers are used on the surface spud-in to ensure a straight surface casing. The bit is usually rotated at about 160 to 180 rpm with 10,000 to 20,000 lb on the bit.

While drilling the surface a wireline survey is generally taken every 500 ft to ensure a straight hole. The shale shaker is also watched to monitor the bit cuttings. On some surface

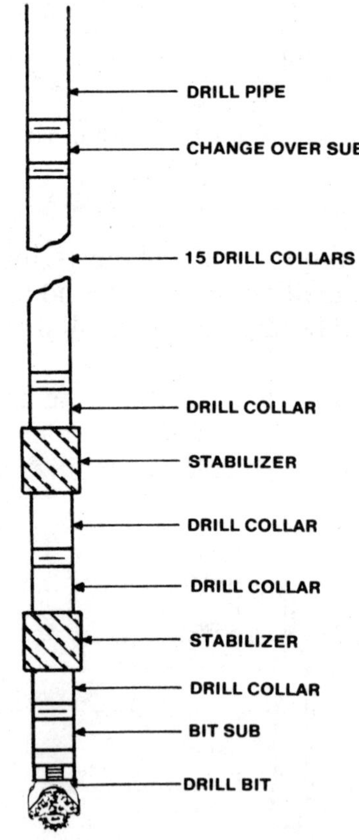

Figure 6-5. The surface bottom hole assembly.

holes pebbles are encountered and are difficult to control. In drilling through pebbles the penetration rate is decreased to allow the mud to clean the hole. You should be sure to add enough gel to bring the cuttings to the surface and keep the pebbles in the wall of the hole.

With proper planning on your part the surface casing will arrive on location, while the surface hole is being drilled. When it arrives the casing must be placed on the pipe rack with the collars facing the V-door. This assures the casing will be picked up correctly. You will need two crew hands to help strap the casing (to strap means to measure) after which you should personally tally the pipe to ensure that all figures are accurate. If the hole is ±3,000 ft, for example, and the total pipe on location is, say, 3,078 ft, then one or two joints are taken out of the string to make it close to 3,000 ft. If one joint is short and one is standard (42 ft) and they add up to 75 ft, both would be eliminated, making the string 3,003 ft in length. The guide shoe and the float collar total 4.62 ft. Added to the string they bring the length to 3,007.62 ft which is perfect for the surface hole.

If the hole is drilled to 3,005 ft and the casing is pulled one foot off the bottom, 3.62 ft will be left above the kelly bushing to allow for the circulation of cement. This is an almost perfect height since it makes setting up the cement manifold easy.

After the hole is drilled to 3,005 ft the casing crew goes to work. You should have notified the casing crew earlier so they will be on standby waiting for the call. You should have also called the cement crew and given them instructions on how much cement and what additives to use. While waiting for the casing (see Figure 6-6) and cement crews to arrive the time should be spent circulating and conditioning the hole and cleaning it up. Build the viscosity up to 50 or 60, while circulating, to keep the surface sands and muds from falling in and bridging over the hole when pulling

Figure 6-6. Surface casing on the rack. (Courtesy of Davenport Horizontal Drilling Consultants, San Antonio, TX)

out for the casing run. In extreme cases of hole sloughing the viscosity could be built up from 80 to 100. To also assure that the hole is in good shape have the crew chain out the hole. (Chaining out the hole keeps the pipe from rotating, which causes parts of the wall to break off and fall to the bottom.)

After tripping out of the hole and once the casing crew is on location, rig up the crew. It normally takes about one hour to rig up the casing crew. It is always wise to ask the casers to bring out an extra pair of tongs in case one breaks down. This will eliminate expensive downtime waiting on tools. Most companies will furnish extra tongs free; if not, the rental fee is around $500. If the casing company refuses to furnish the extra tongs, do not use that company on the next job.

After rigging up to run the casing pick up the first joint and apply lock putty to the threads. This will bond the threads together like a weld. Install the guide shoe to the first joint, and run it about halfway through the floor. Then add a centralizer and use a stop ring to hold the centralizer in place, so it will not slip. Run the joint through the floor and install a centralizer every other collar, until three are used. Some engineers will want you to use more or less, so the number of centralizers used really depends on the engineer in charge of the well. (See Figure 6-7.)

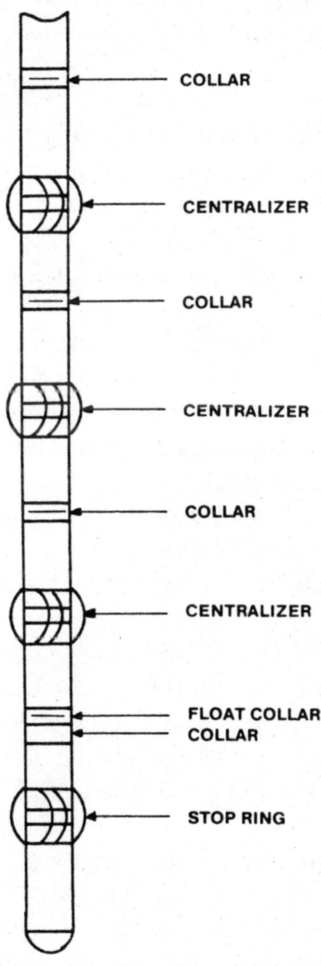

Figure 6-7. Example of the use of centralizers in a drill string.

Run the casing in the hole, filling it every three to five joints with fluid to counter the buoyant effect and to keep the casing from crushing under hydrostatic pressure. Add two centralizers to center the casing in the conductor pipe (Gulf Coast drilling). A cement basket can be installed about 5 ft from the bottom of the conductor to help reduce the hydrostatic head of the cement if it seals. (See Figure 6-8.) I believe that the cement basket is not necessary, but always remember that you do not have "downhole glasses," so do what the engineer says to do on the prognosis.

If a problem is encountered in tagging the bottom, because of fill or hole conditions, the cement manifold can be installed and the casing can be washed down to total depth (TD) by breaking circulation and working the pipe down. Then additional joints can be added as the casing is washed down to TD.

If the casing will not wash down, the hole is probably not as deep as requested. At this point there is no choice but to cement the casing right there. This puts the cement manifold higher up and causes problems for the crews on installation. When cementing, the collars must be below ground level, or at least 5 to 6 ft above ground level. This is important when nippling up.

If the casing does not touch bottom with the allotted pipe, it means the hole is too deep. One of the joints previously set aside must be added to the string. Tag bottom first, then decide whether to have a cement plug below the shoe or to be 1 ft off bottom. If the hole is deeper than 3 ft, you need to leave the joint in the string and be 1 ft off bottom. If it is less than 3 ft, lay the joint down and cement the string where it is.

Consult the cement engineer, and make sure he brought out the right cement combination. Mistakes can be made and it is important to run the type of cement the operator requested into the hole. It is also wise to sit down with the cement engineer and go over his calculations for accuracy.

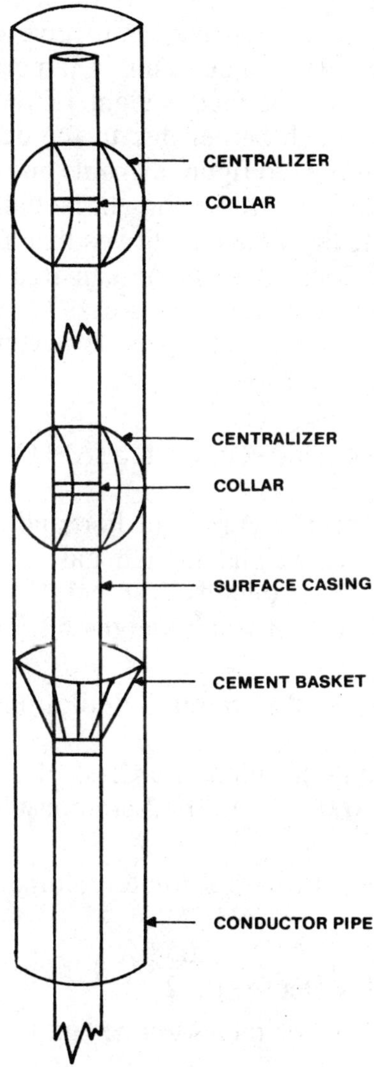

Figure 6-8. Surface casing with a cement basket for reducing the hydrostatic head of the cement.

All cement jobs will have a lead and a tail cement. The lead cement will be light weight so it can be pumped more easily. The tail cement is heavier to give a good cement plug around the casing shoe. This is important for testing the shoe for equivalent mud weight (EMW). The heavy cement will dry much harder due to the density. For purposes of learning how to figure cement, one combination of cement instead of two will be shown. Not all cement in this book will include the simple additives that are used in most cement combinations. If a person can figure the basic cement, then adding the additives is easy.

The cement calculations are simple if you remember the following facts.

1. A sack of cement weighs 94 lb (API Classes A through H).
2. Absolute specific gravity of Portland cement is 3.14.
3. Water always weighs 8.33 lb/gal.
4. One cubic foot contains 7.48 gal.
5. One cubic foot of water weighs 62.3 lb.

Typical Cement Calculations

In the following calculations, Class H cement and 48% water are used (0.0382 and 0.12 are constants):

```
components wt. (lb)  ×  absolute volume  =  volume (gal)
                          (gal/lb)
94 lb cement             × 0.0382            = 3.59 gal
45.12 lb 48% water  × 0.12                   = 5.41 gal
                          total volume (gal)  = 9.00 gal
```

total components' wt. = 94 lb cement + 45.12 lb water
= 139.12 lb

The slurry density formula is:

$$\text{slurry density} = \frac{\text{total components' weight (lb)}}{\text{total components' volume (gal)}}$$

$$\text{slurry density} = \frac{139.12}{9.00} = 15.45 \text{ lb/gal}$$

The slurry yield formula is:

$$\text{slurry yield} = \frac{\text{total components' volume (gal)}}{7.48 \text{ gal/ft}^3}$$

$$\text{slurry yield} = \frac{9.00 \text{ gal}}{7.48 \text{ gal/ft}^3}$$

$$\text{slurry yield} = 1.20 \text{ ft}^3/\text{sk}$$

Now that the slurry density and the slurry yield are known, the next step is to look in the cement book (such books are provided by cement suppliers/vendors) under volume and height between casing and hole. Assume a hole diameter 14¾ in. and a casing OD of 10¾ ft. Then using the chart and obtaining cubic feet per linear feet the number is 0.5563. Multiply by height in feet.

Assume the hole is 3,004 ft deep; then to find cubic feet use the formula:

$$0.5563 \times 3{,}004 = 1{,}671.12 \text{ ft}^3 \text{ of volume needed to fill the annulus from top to bottom of hole.}$$

Now to find the number of sacks needed the formula is:

$$\text{sacks} = \frac{\text{annular volume ft}^3}{\text{slurry yield ft}^3/\text{sack}}$$

$$\text{sacks} = \frac{1{,}671.12 \text{ ft}^3}{1.20 \text{ ft}^3/\text{sk}}$$

$$\text{sacks} = 1{,}392.6$$

So it takes 1,392.6 sacks to fill the annulus.

In the oilfield cement is always ordered in *sacks* instead of cubic feet. In other words order 1,392 sacks of Class H with 48% water, not 1,671.2 ft^3.

On the surface cement job always figure 100% excess to ensure a good cement job. The 100% is to fill all washouts, to get a good bond downhole, and to get the cement to come to the surface. If there is no return to the surface with the 100% excess, then there are two possible problems:

1. The washouts were worse than expected.
2. The formation broke down and the cement went south, as we say in the oil business.

If this happens, do not panic—a 1-in. job normally does the trick. Run 1-in. tubing down the annulus between the surface casing and conductor pipe, and try to tag the cement. Then pump cement down the hole until good returns come back to the surface. Pull the tubing out and allow the cement to set (see Figure 6-9). Sometimes the tubing will not pull out. In such cases it is simply cut off and left in the hole with no ill effects to the cement job.

To figure cement at 100% excess, it is necessary to multiply by 2, so in our example, 1,392 sacks × 2 = 2,784 sacks. In the following calculation 0.1781 is a constant and

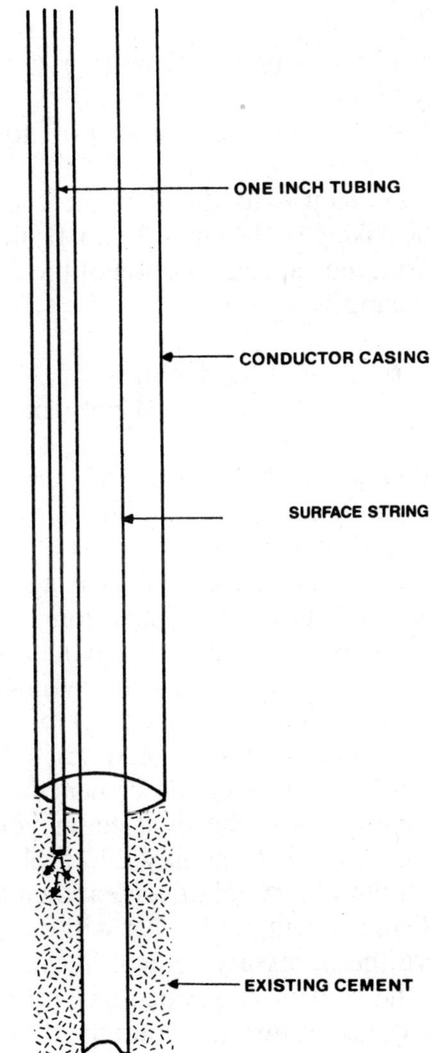

Figure 6-9. Diagram of a 1-in. job.

3,342 is obtained by multiplying 1,671 ft³ by 2 to equal 100% excess.

slurry bbl = slurry volume (ft³) × 0.1781
slurry bbl = 3,342 ft³ × 0.1781
slurry bbl = 595.21 (rounded off to 595 bbl)

The next step is to calculate the capacity of the casing. Since the casing is 10¾ in. 40.5 lb/ft, the capacity is 0.0981 bbl/ft from the capacity section of the vendor cement book. The casing is:

3,007.62 ft × 0.0981 bbl/ft
= 295.04 bbl (rounded off to 295 bbl)

We now have the number of barrels it would take to displace all the cement out of the casing and into the annulus with mud; however, since there is a float collar, some cement will need to remain in the casing for testing purposes. The capacity must be calculated from the bottom of the float collar (back pressure valve) to the bottom of the guide shoe, which for our example we will say is 45 ft.

By the preceding formula 45 ft × 0.0981 = 4.41 bbl and by rounding off this becomes 4.5 bbl. So to pump enough mud to displace the cement in the casing and still leave cement below the float collar you would pump 295 bbl minus 4.5 bbl, or simply 290.5 bbl.

So first pump 595 bbl of cement then follow it with 290.5 bbl of displacement mud. That will put the cement in place and leave the necessary cement in the casing to test.

After the cement is pumped check for backflow to see if the float collar or insert is holding. If not, simply pump the reclaimed mud back down the hole and close the cement manifold at the surface.

There are two types of float collars:

1. Standard
2. Automatic fill

The automatic fill is better but is much more expensive. If the float collar is used after the cement is pumped and a

Figure 6-10. The surface job is finished. (Courtesy of Davenport Horizontal Drilling Consultants, San Antonio, TX)

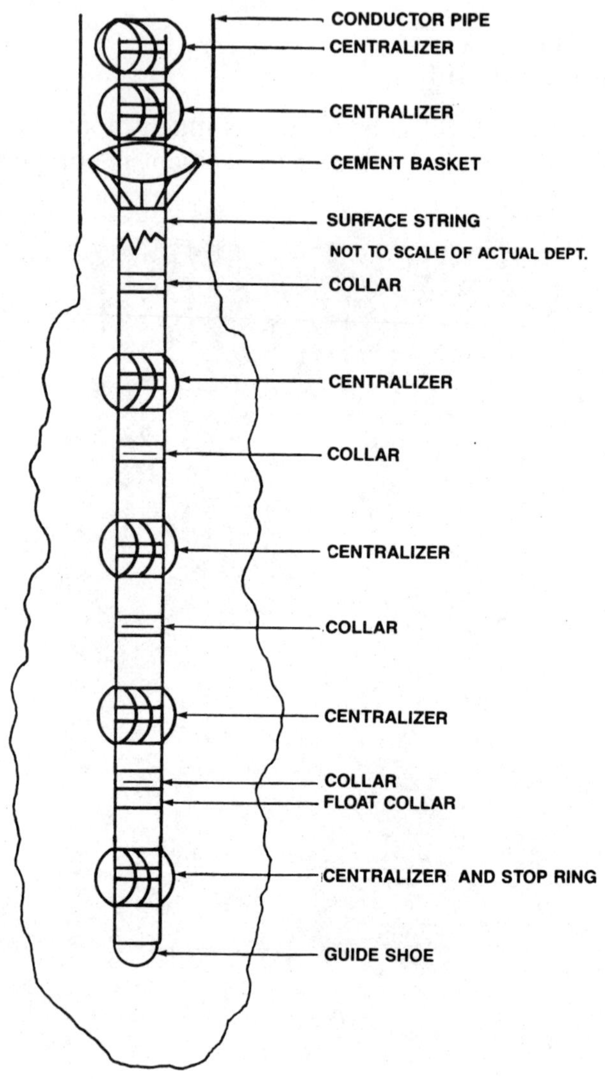

Figure 6-11. The surface string cemented in the hole.

plug placed after the cement, pump it down the hole with the 290.5 bbl. When the plug lands on the float collar the pressure will rise at the surface. This is called "bumping the plug." Record this pressure in your records.

Some operators like to pump a plug in front of the cement to help isolate the mud from the cement and to clean the casing. When the plug hits the float collar, pressure is increased and the plug breaks down, allowing the cement to flow through the float collar. The second plug will not break down and is pumped after the cement is pumped. It isolates the cement from the mud and also cleans the casing wall as it goes down the casing.

The surface casing is now cemented (see Figures 6-10 and 6-11). So now just wait and let it set. Normally the

Figure 6-12. Horizontal well. (Courtesy of Davenport Horizontal Drilling Consultants, San Antonio, TX)

surface should set for 8 to 12 hours. If more operators would wait 24 to 30 hours, it would eliminate having to squeeze the casing seat, in most cases. The prognosis will tell you how long to wait.

(**H**) In horizontal wells (Figure 6-12), the surface casing is normally not new, but used. Some of the casing that I have buried looked as if it came out of a river, but it worked well because of the depth and the lack of pressure. The float equipment is basically the same setup, except that an insert is used instead of a float collar.

7
Nippling Up the BOP Stack

After the cement sets the next step is to cut the casing. Release the weight of the casing from the blocks and disconnect the cement manifold. Keep the blocks on the casing to remove the top of the casing after it is cut.

Have the welder measure the bradenhead or casing head, and have him cut off the surface casing so that the flange on the casing head will be at ground level. He will bevel the surface casing to improve the weld to the casing head. After he welds the head, as it is called in the field, let it cool for one or two hours in cold weather or for half an hour in warm weather.

Normally the head is tested to 1,000 psi for 15 minutes—if it is going to leak it will do so in 1 or 2 minutes. Check the prognosis for the correct test time. Most leaks are pin holes and are easily repaired. After a leak is repaired the head must be retested, to ensure there are no leaks. Most well head companies send a service man to supervise the installation. Once he knows what is required he can do the

Figure 7-1. Nippling up the stack. (Courtesy of Davenport Horizontal Drilling Consultants, San Antonio, TX)

job. Never let the hired welder test his own welds. Always let the service man do that to avoid a conflict of interest.

Nippling up is one of the most important parts of drilling a well (Figure 7-1). If the BOP cannot handle a solid "kick" the well may be lost. The proper way of nippling up is a debatable issue, but you should always take into account the problems in fighting kicks.

The only way to nipple up and be safe is to use a three-ram system. The system from the bradenhead has a pipe ram, choke, kill line, another pipe ram, the blind ram, and then the annular preventer. On top of this stack is the bell nipple (see Figure 7-2).

With this system any kick can be handled with no problem. If the annular preventer goes out, the well can still

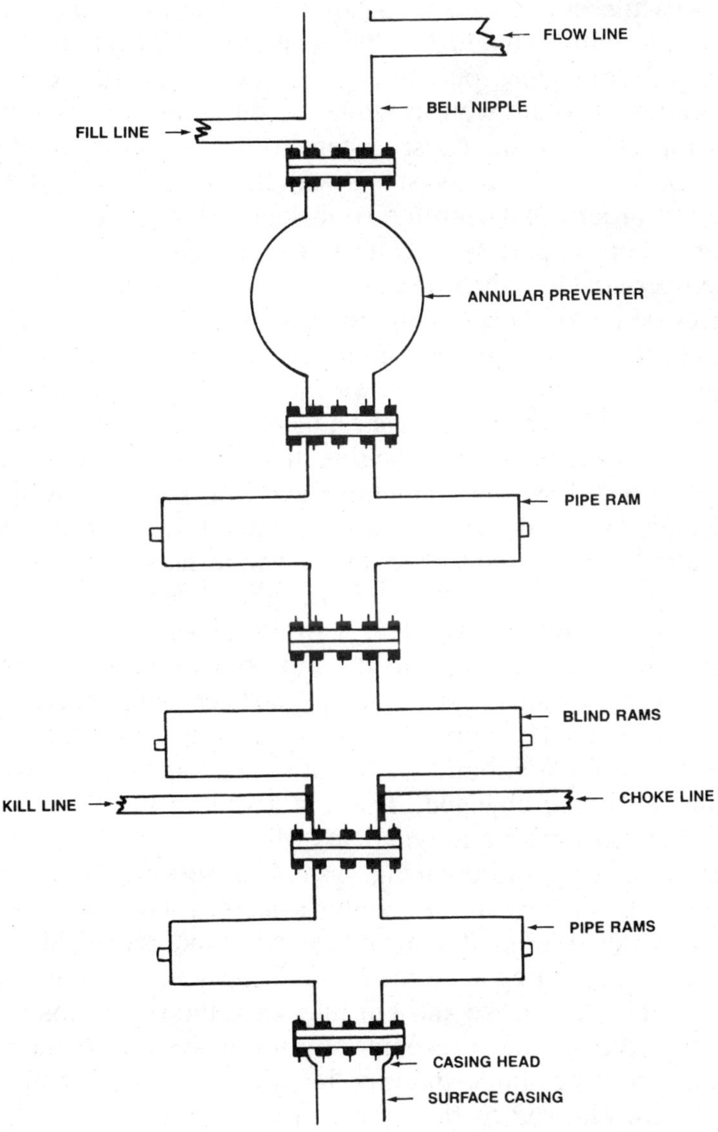

Figure 7-2. The safest blowout system uses three rams.

flow while killing operations are under way, by closing the top pipe ram (explained in the chapter on blowouts). If the top pipe ram goes out, then the bottom pipe ram can be closed to shut the well in while the pipe rams are changed on top. This is the safest system to use, but very seldom do you find three-ram systems with the rig. Most engineers do not order them with the rig because they do not understand their importance. Unless the consultant explains to the operator why the three rams are needed he will in most cases be stuck with a two-ram system.

There have been many arguments in the field about the right way to nipple up a two-ram stack. Remember the annular preventer is the weakest part of the blowout system. In most cases the setup begins at the bradenhead with a pipe ram followed by the choke and kill line, with a blind ram on top, and the annular preventer following it. This system is considered to have two advantages: since a surface shoe can hold very little pressure, the shoe would go out before the annular preventer. Also, if the well kicks and the annular preventer goes out the pipe rams could be closed and the blind rams pulled out and replaced with another set of pipe rams. The only thing wrong with this theory is that the pressure could build up and blow out the shoe while the rams are being changed. There is also the possibility that the bottom pipe rams will leak while the other rams are being changed. Although this system has serious flaws, 90% of the rigs drilling today use this system. (See Figure 7-3.)

Now let's look at the system with the rams changed around. Beginning with the bradenhead we have the blind rams, then the choke and kill line, then the pipe rams, and on top, the annular preventer. With this system, if the annular preventer fails, the pipe rams can be closed and the well still allowed to flow through the choke while a third ram is installed in place of the annular preventer. The big problem with this system is if the choke valves cut out. The

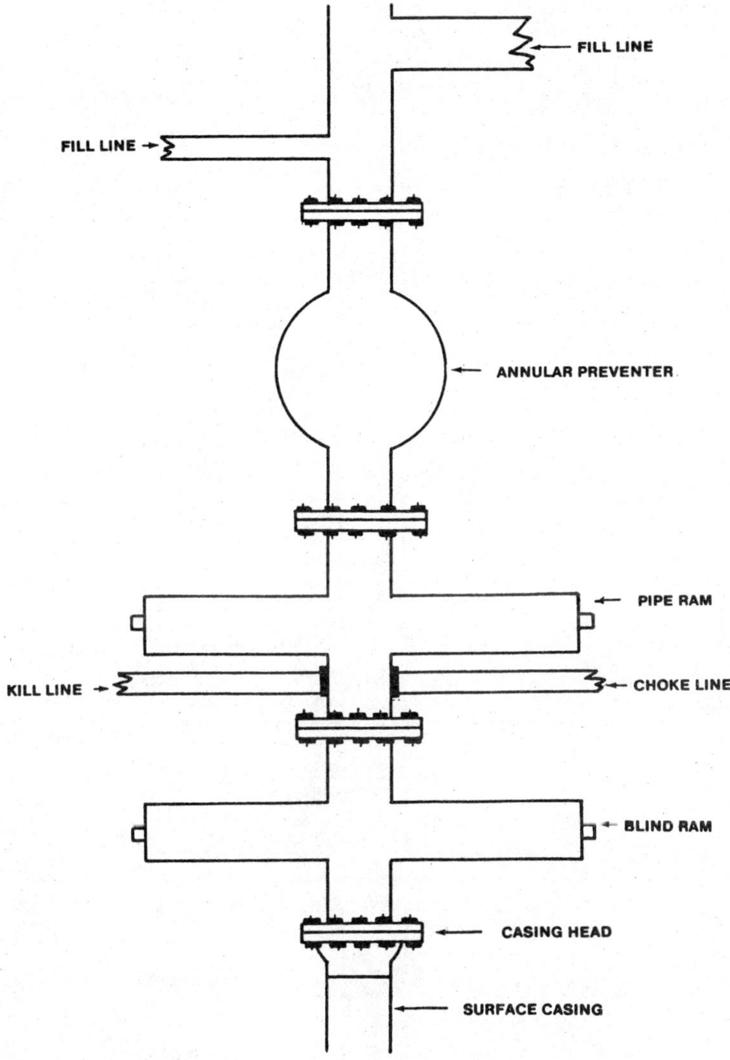

Figure 7-3. The unsafe two-ram blowout prevention method.

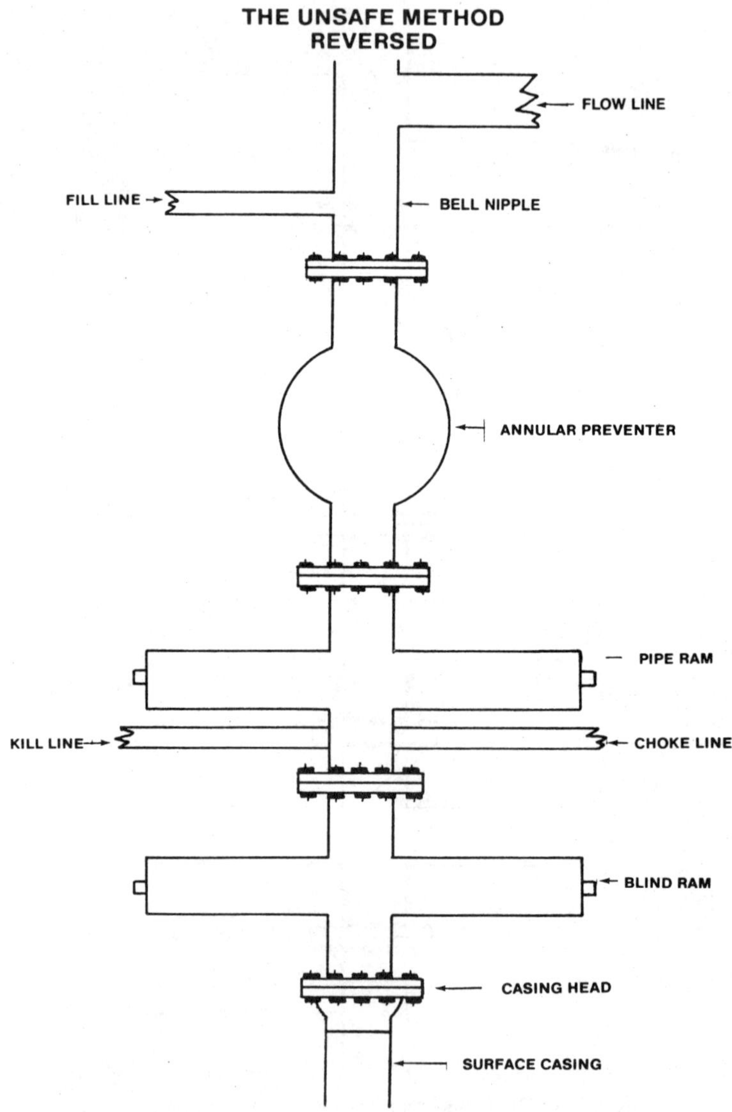

Figure 7-4. The unsafe two-ram method reversed.

well is exposed to the world and fire could easily break out. (See Figure 7-4.) I cannot recommend either two-ram method, since both have unsafe features. To be safe a three-ram system must be used.

As the oilfield gets more specialized so do the services. When nippling up, you should try to find a testing company that has a hydraulic wrench service to tighten the stack. The fee is usually around $350 per hour. If they are fast the operator saves money.

To test the stack a portable tester is needed. In most cases there are service companies to perform testing. They will furnish the required equipment and the test plug. The test plug is a device that is run into the hole on a joint of drill pipe. It seals in the bradenhead so the tester can pressure

Figure 7-5. Pressure testing the blowout preventers. (Courtesy of Wild Bunch Hellfighters, San Antonio, TX)

up on the stack for the tests. The following lists areas to be tested and the pressure to which you should test them:

Area	Pressure (psi)
BOP	5,000
Annular preventer	2,500
Choke manifold and superchoke	5,000
All lines to the choke manifold	5,000
HCR valve	5,000
All TIW valves	5,000

(TIW is a trademark of Texas Iron Works)

Some companies require different test pressures. Check with the oil company engineer since deeper wells require higher test pressures than do shallow wells.

Pressure testing should be performed every seven days to be sure the blowout equipment is ready at all times (Figure 7-5). The vibration of the drilling rig loosens the chain nuts

(continued on next page)

Figure 7-6. A ram-type blowout preventer. (Courtesy of Hydril Co.)

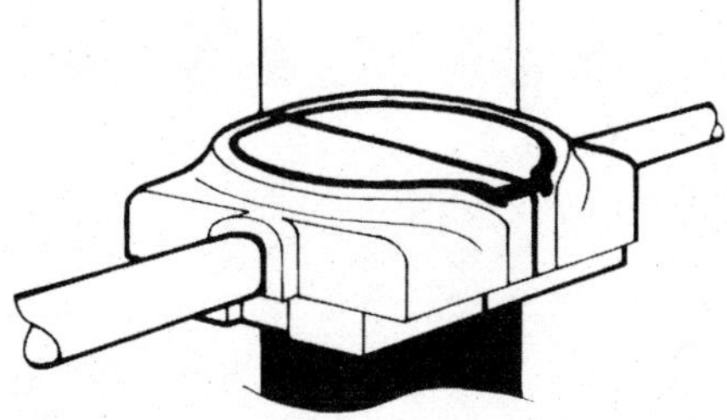

BLIND RAM CLOSED

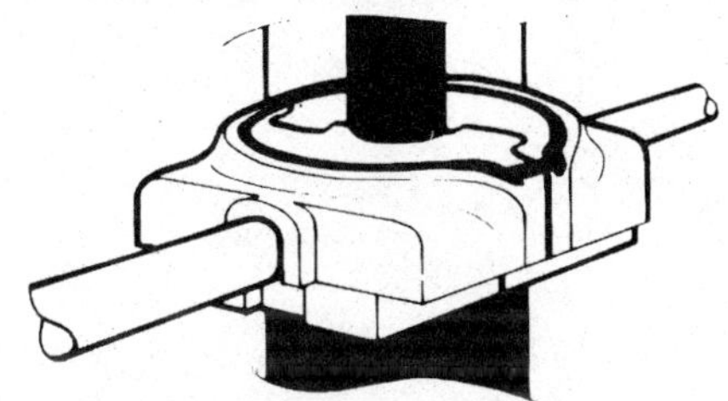

PIPE RAM CLOSED ON DRILL PIPE

Figure 7-6. Continued.

on the stack, but checking every seven days is sufficient to ensure a safe operation.

When pipe rams are in operation they close around the drill pipe and seal off the annulus of the well. Blind rams close and seal off the open hole. When drill pipe is pulled out of the hole, make sure that the blind rams are shut to keep objects from falling down the hole, creating a fishing job. Rams are shown in Figure 7-6.

When the nippling-up job is over, it is important to install a wear ring in the bradenhead so the kelly's turning will not wear out the bradenhead.

8
Setting Up the Bottom Hole Assembly

A good "bottom hole assembly" (BHA) is essential to drilling, whether drilling is straight or directional. This text discusses vertical BHAs and horizontal BHAs. The drilling consultant decides how to run the BHA on vertical wells, but on directional wells it is up to the directional driller. Figure 8-1 shows the BHA for a straight hole. The next few pages discuss straight hole drilling.

The shock sub is a new innovation to the oilfield, and some consultants and engineers will not use them. However, shock subs serve the same purpose as shock absorbers on a car. Shock absorbers keep the tires on the road and reduce tire wear. In drilling the shock sub keeps the bit on the bottom of the hole and reduces bouncing. If the drill string bounces, the threads on the drill collars may be damaged. The bit will wear out sooner, and damage to drill pipe is possible. A 3- or 4-in. bounce on the bottom is hard on the string, just as it would be on passengers riding in a car with no shocks.

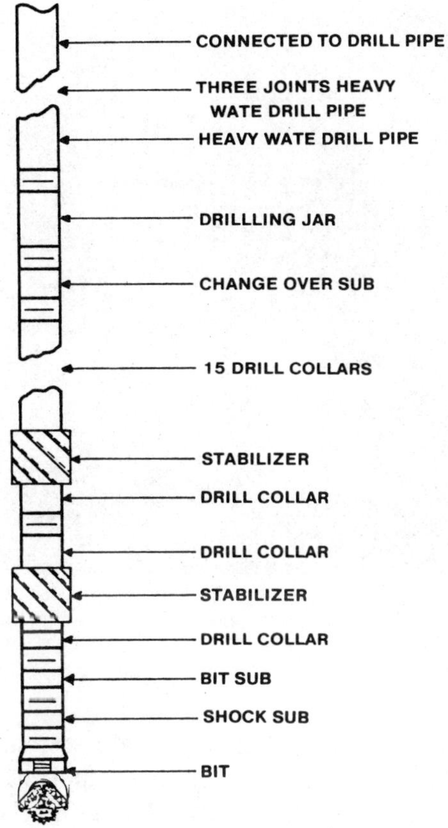

Figure 8-1. Example of a bottom hole assembly.

There are many shock subs on the market, and there are many fine companies with excellent subs. It is a rental item. Check the rotating hours it is rated for and check after each trip for loss of strength. If the shock sub is used the bit sub will go on top of it instead of on the bit to sub it into the drill string.

A drill collar (see Figure 8-2) follows the bit sub, and a stabilizer follows that. A stabilizer is designed to keep the

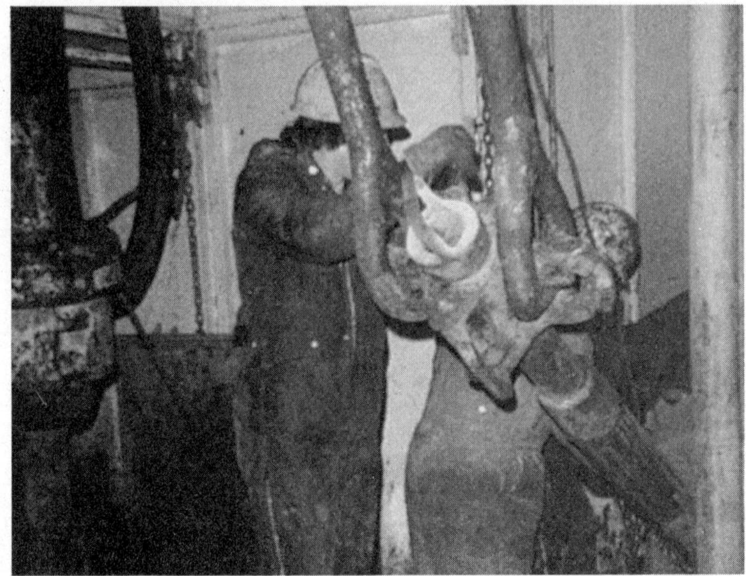

Figure 8-2. Picking up drill collars. (Courtesy of Davenport Horizontal Drilling Consultants, San Antonio, TX)

string stable in the hole. Call for a stabilizer salesperson to help design the BHA. Stabilizers are much needed on the Gulf Coast because of the sands, shales, and soft drilling. There are several types of stabilizer designs. They are:

- Welded blade
- I.B.S.
- Insert types

They also come in straight or spiral blade. The welded blade unit has some problems because the blades can break off, causing a fishing job. The I.B.S. stabilizer cannot break off, but it is expensive to run. Your salesperson will help assist you in deciding which type to use in the hole. In hard rock areas stabilizers are not as important, unless the well is in crooked hole country. The prognosis will tell you when to use stabilizers.

Follow the stabilizer with two drill collars and another stabilizer, placing the stabilizers at 30 feet and 90 feet above the bit. When you are adding stabilizers to the string, figure

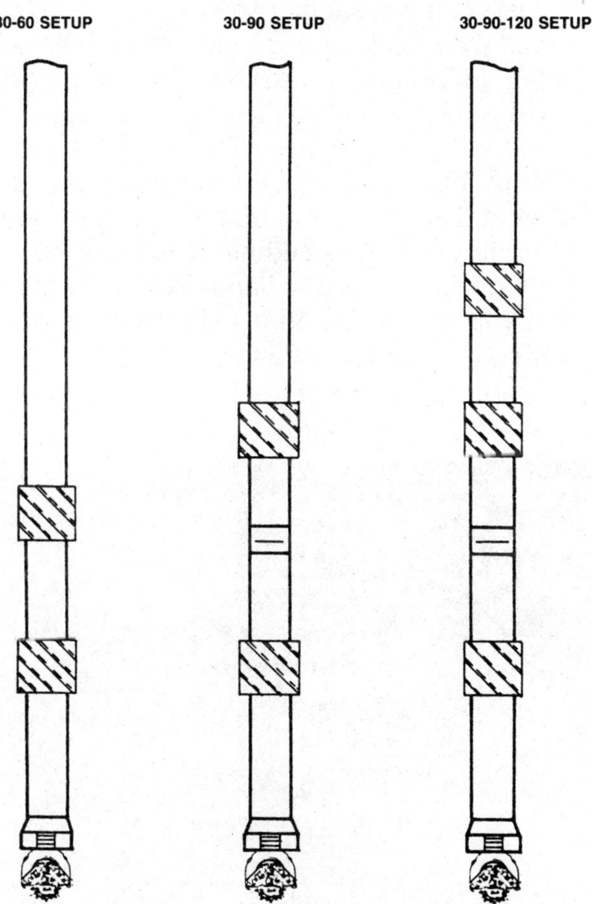

Figure 8-3. Bottom hole assembly setups.

placing the drill collars 30 ft apart and putting a stabilizer at 30 ft and 60 ft. That means the first stabilizer is placed one collar above the bit, one more collar is added, and, then another stabilizer is placed on top of it. So you have a 30-60 setup. A 30-90 or 30-90-120 setup is figured in the same manner (see Figure 8-3). Determine how many collars are needed for weight, and add them to the string.

Drill collars come in slick or spiral designs. Spiral collars are preferred over slick because they reduce the possibility of becoming differential stuck. Unfortunately, most rigs are running with slick collars, so the consultant has to live with them.

Some drill collars have a slip indention cut so that the slips will hold them on the rotary table. If they do not have the cut, a safety clamp (wedding band) is used whenever tripping. No driller or floor hand likes to use the safety clamp, but you must insist on it. Always stand on the floor if the drill collars are not indented to make sure the safety

Figure 8-4. Drilling jars. (Courtesy of Dailey Petroleum Services Corp.)

clamp is used. The clamp will catch the collars and keep them from going to the bottom, thus preventing a fishing job and lost rig time.

The next item is the drilling jar (see Figure 8-4). This is designed to jar the pipe when the pipe gets stuck. The jar will move the pipe in most cases, causing it to break free. Usually a changeover sub (abbreviated XO) is needed to

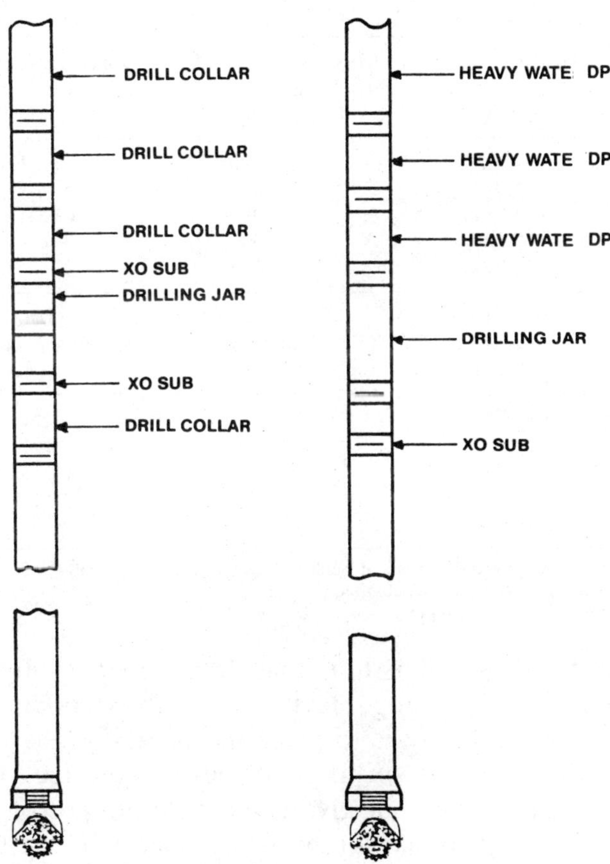

Figure 8-5. Methods to sub in drilling jars.

TABLE 8-1.   L.I. Rotary Drilling Jar Specifications.

Outside dia. (in.)	Inside dia. (in.)	Tool joint size (in.)	Tensile yield (a) (lbs)
4⅛	1⅛	2⅞ A.P.I. I.F.	396,000
4¾	2	3½ A.P.I. I.F.	449,000
6¼	2¼	4½ A.P.I. I.F.	751,000
	2¼	4½ A.P.I. X.H.	751,000
6⅞	2½	5½ A.P.I. Reg.	1,045,000
7¾	2¾	6⅝ A.P.I. Reg.	1,148,000

Outside dia. (in.)	Housing torque at assemble (ft/lbs)	Max. recommended drilling torque (ft/lb)	Max. recommended drill collar size below jar
4⅛	7,600	4,500	4½″ OD
4¾	12,000	5,500	5½″ OD
6¼	18,000	11,000	6¾″ OD
7⅛	22,500	14,000	7¾″ OD
7¾	26,000	18,000	9″ OD

Outside dia. (in.)	Free travel upstroke (in.)	Free travel downstroke (in.)	Total stroke (in.)
4⅛	7⅞	8	15⅞
4¾	8⅝	8	16⅝
6¼	7⅛	8	15⅛
6⅞	7⅞	8	15⅞
7¾	7¾	8¼	16

(a) The tensile yield and the torsional yield values are calculated per API RP 7G based on nominal dimensions and the published yield strength of the material and do not constitute a guarantee, actual or implied.

go from the drill collars to the jar. Some people sub the jar into the drill collar string, leaving two or three collars above the jar. The best way is to place the jar above the collars and then run five or six joints of heavy wate (see Figure 8-5) drill pipe. This will give enough hitting power to get the full effect of the jar on the string. Normally set the jar to go off at 80,000 lb over the string weight. It can, of

Torsional yield (a) (lbs)	Normal jar settings		Maximum allowable overpull (lbs)
	Upstroke (lbs)	Downstroke (lbs)	
18,000	53,300	28,200	58,000
30,000	69,700	32,900	75,000
58,200	94,300	37,600	118,000
58,200	94,300	37,600	118,000
84,000	98,400	42,300	120,000
120,300	102,500	42,300	125,000

Max. recommended drill collar size above jar	Approx. length extended (ft)	Approx. weight (lbs)
4⅛" OD	26½	820
4¾" OD	30½	1,340
6¼" OD	34	2,500
6¾" OD	34½	3,150
7¾" OD	35½	4,000

Maximum circulating pressure	Maximum hydrostatic pressure (b)	Maximum BHT (c)
5,000 psi	None	325°F
5,000 psi	None	325°F
5,000 psi	None	325°F
9,000 psi	None	325°F
5,000 psi	None	325°F

(b) In cases of high mud weight & low BHT or low mud weight & high BHT, consult your salesperson.
(c) Hot hole packing is available for specific applications.

course, be set at any weight before it goes off. See Tables 8-1 and 8-2 for specifications of two types of drilling jars.

The bottom hole assembly can be varied to fit the job, but remember that the main purpose of the BHA is to furnish weight to the bit and control the hole. Before deciding how many drill collars to put in the string, determine the max-

TABLE 8-2.   Dailey Hydraulic Drilling Jar Specifications.

Outside dia. (in.)	Inside dia. (in.)	Tool joint size (in.)	Tensile yield (a) (lbs)
3¾	1¾	2⅜ A.P.I. I.F.	288,000
3¾	1.937	2⅜ EUE	288,000
4¼	1.937	2⅞ A.P.I. I.F.	336,000
4¾	2¹⁄₁₆	3½ A.P.I. I.F.	436,000
6¼	2¼	4½ A.P.I. I.F.	832,000
	2¼	4½ A.P.I. X.H.	832,000

Outside dia. (in.)	Housing torque at assemble (ft/lbs)	Max. recommended drilling torque (ft/lb)	Max. recommended drill collar size below jar
3¾	3,000	2,250	4¼" OD
3¾	3,000	2,250	4¼" OD
4¼	6,000	4,500	4¾" OD
4¾	12,000	8,000	5½" OD
6¼	18,000	12,000	6¾" OD

Outside dia. (in.)	Free travel upstroke (in.)	Free travel downstroke (in.)	Total stroke (in.)
3¾	4¼	16 (b)	22
3¾	4¼	16 (b)	22
4¼	4½	18 (b)	24¼
4¾	5	5	13½
6¼	6¼	6¼	16½

(a) The tensile yield and the torsional yield values are calculated per API RP 7G based on nominal dimensions and the published yield strength of the material and do not constitute a guarantee, actual or implied.
(b) The smaller sizes jar up and bump down.
(c) Hot hole packing is available for specific applications.

imum weight to be run on the bit, then add 25% more collar weight.

The pipe and the BHA have a buoyant factor of about 20%, plus a 5% safety factor, which equals 25% added weight.

Torsional yield (a) (ft/lbs)	Minimum overpull		Maximum allowable overpull	
	Up (lbs)	Down (lbs)	Up (lbs)	Down (lbs)
10,600	None	N/A (b)	44,000	N/A (b)
10,600	None	N/A (b)	44,000	N/A (b)
16,400	None	N/A (b)	66,000	N/A (b)
21,200	None	None	95,000	95,000
49,300	None	None	200,000	200,000
49,300	None	None	200,000	200,000

Max. recommended drill collar size above jar	Approx. length extended (ft)	Approx. weight (lbs)
3¾" OD	22	475
3¾" OD	20	460
4¼" OD	24½	500
4¾" OD	32	1,200
6¼" OD	33	2,050

Maximum circulating pressure	Maximum hydrostatic pressure	Maximum BHT (c)
5,000 psi	None	400°F
5,000 psi	None	400°F
5,000 psi	None	400°F
5,000 psi	None	400°F
5,000 psi	None	400°F

Example

If the maximum weight on the bit is 35,000 lb, then add 25%, which is 8,750 lb.

8,750 + 35,000 = 43,750 lb of collars

Collar weight is measured by the foot rather than by the joint. Collar weight is listed in the collar book provided by the vendor in pounds per foot. To find the amount of feet of drill collars needed divide:

$$\frac{43,750 \text{ lb of collars}}{75 \text{ lb/ft}} = 583 \text{ ft of collars needed}$$

To find the number of collar needed, divide the feet needed by the average length of the collars.

Example:

$$\frac{583 \text{ ft of collars needed}}{30 \text{ ft average length of DC}} = \begin{array}{l} 19.43 \text{ (rounded off} \\ \text{to 20 drill collars} \\ \text{needed to run 35,000} \\ \text{lb of the drill bit)} \end{array}$$

Remember, the drill collars furnish weight to the bit; the drill pipe does the rotating work.

The bottom hole assembly can be changed in many ways. If the BHA is designed similar to the one in Figure 8-3, you will be able to drill a straight hole faster than with most assemblies (with some adjustment to weight and bit RPM for the formation drilled). A good BHA will make the consultant's job easier and more trouble free.

Always strap each drill collar and check its OD and ID and check all parts of the BHA. This information should be entered on a BHA work sheet as part of your records. The information is valuable in case a fishing job becomes necessary.

Remember that when you are in trouble, having your house in order will help keep you from getting fired. *Keep an accurate account of everything you do at all times.* (See Figure 8-6 for an example of a form for recording bottom hole assemblies.)

(**H**) In horizontal drilling, after the vertical hole is drilled, the world can turn upside down. Most directional drillers like the stronger S-135 drill pipe because it is more successful. The drill pipe goes in first after the directional tools,

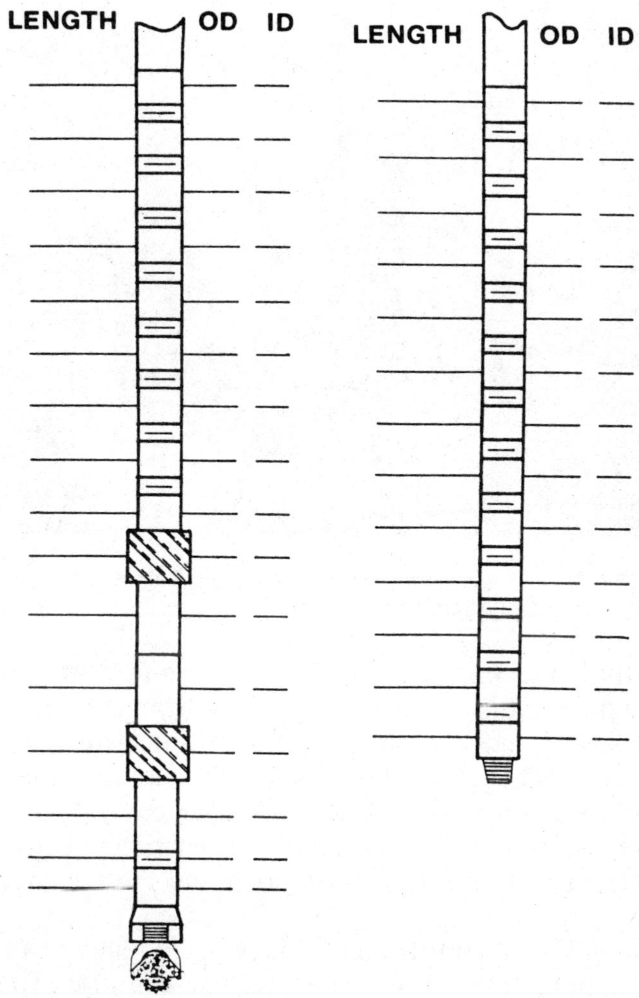

Figure 8-6. Form for recording length, ID, and OD of bottom hole assemblies.

and then come the heavy wate drill pipe, the drill collars, the jar, and the drill pipe (see Figure 8-7). That may seem nightmarish, but it is the standard setup. Some directional drillers want to run only drill pipe; personally I like to hang a jar and drill collars in the hole in case the string gets stuck. Mainly make sure that the directional motor is locked down

Figure 8-7. Roughnecks tagging bottom. (Courtesy of Davenport Horizontal Drilling Consultants, San Antonio, TX)

with thread lock. The number one reason for losing motors in the hole is that the motors unscrew from the BHA. This is caused by the whiplash effect of the pipe rotating, stopping, sliding, and moving. Most directional drillers will thread-lock if you ask. If not, call your company engineer before the pipe goes in the hole. Remember if you have trouble, the clock keeps on running and the operator pays for your mistakes.

In the BHA on horizontal wells, you are really in a Catch-22 situation; either way you move, you can lose. The only winner is the directional driller and company, so try to run drill collars (DCs) and a jar and lock down the tools with thread lock. That should ensure some measure of success.

The directional driller will tally all the pipe in the hole by ID and OD and length. However, a consultant should make sure he also has a tally with his records in his own trailer in case of problems downhole.

9
Drilling Out Surface Casing, Tests, and Squeeze Jobs

Before the drilling operation can get underway again, the surface casing must be drilled out and tested 10 ft below the shoe (Gulf Coast area). To do this, strap and record the depth that the cement is tagged. Then close the annular preventer and pressure up on the casing with the rig pump through the drill string. Pressure up to 1,500 psi or whatever the prognosis calls for. Check the casing for leaks. If there are no leaks, proceed with the next step. If the casing leaks, call for a squeeze tool and find the leak, because a cement job will be necessary. In this section we assume the test passed with no leaks.

Next drill out the float collar and cement 10 ft above the guide shoe. Test the casing to 1,500 psi again to check for leaks in the float collar threads. Drill out the shoe and 5 ft of the new formation (see Figure 9-1). Check the prognosis to determine what *equivalent mud weight* (EMW) to pressure up on the shoe. The EMW is the predetermined test pressure based on the geology to which the operator wants to test

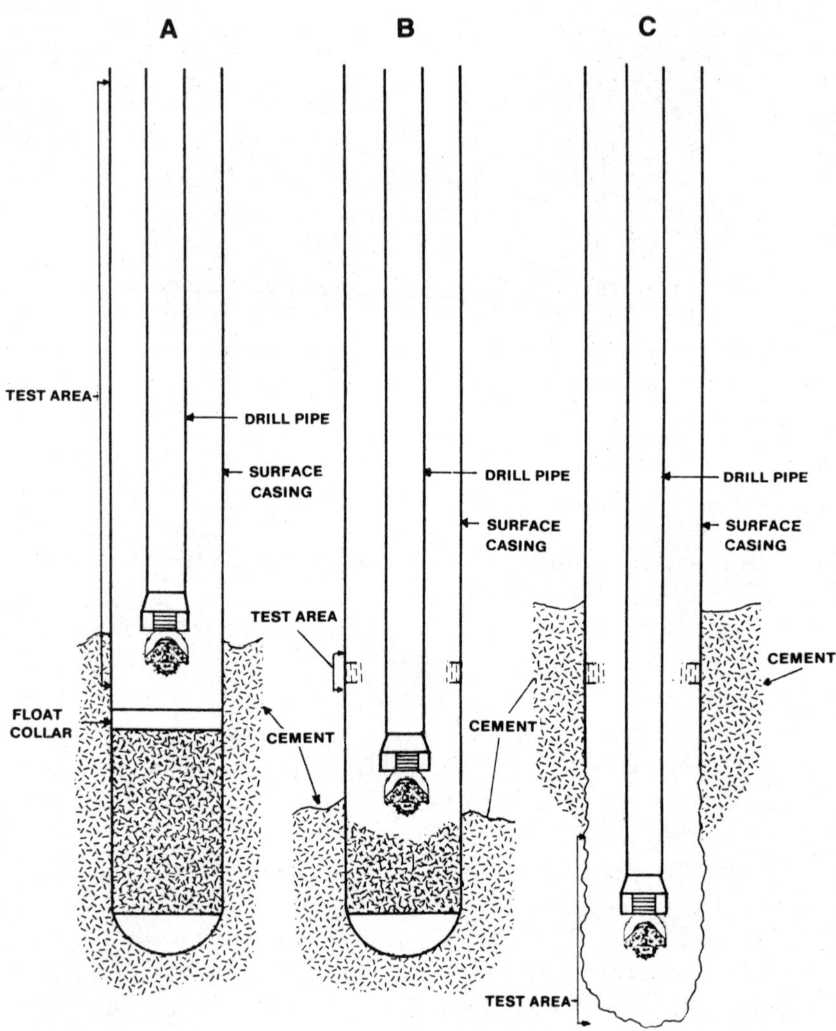

Figure 9-1. Method of testing (A) the casing, (B) the float collar, and (C) the formation.

the casing shoe. The test is given if the well is going to encounter high-pressure gas. On the Gulf Coast normally a 13.5 ppg (pounds per gallon) mud weight is used to test the surface shoe. The EMW is what is needed at the shoe depth, before drilling begins. The operator will put on the prognosis the shoe test pressure. If the shoe is tested and holds, then if the mud weight needs to be raised to the weight of 13.5 ppg, the shoe will hold and not leak (see Figure 9-2). To find the EMW, first find the existing mud weight in the tanks. Then do the following calculation (0.052 is the constant for converting ppg to psi):

depth $\times$ mud weight (ppg) $\times$ 0.052 = BHP

Example:

3,014 ft (depth) $\times$ 8.9 ppg (mud weight) $\times$ 0.052
$$= 1{,}394 \text{ psi}$$

Next find the hydrostatic pressure on the EMW, as listed in the prognosis and do the following calculation:

depth $\times$ EMW $\times$ 0.52 = psi

Example:

3,014 ft (depth) $\times$ 13.5 ppg (EMW) $\times$ 0.52
$$= 2{,}115 \text{ psi}$$

Now subtract existing pressure from the test pressure of EMW:

2,115 − 1,394 = 721 psi

So 721 psi will be needed at the surface to get the EMW of 13.5 ppg mud. Pressure up on the shoe by closing the

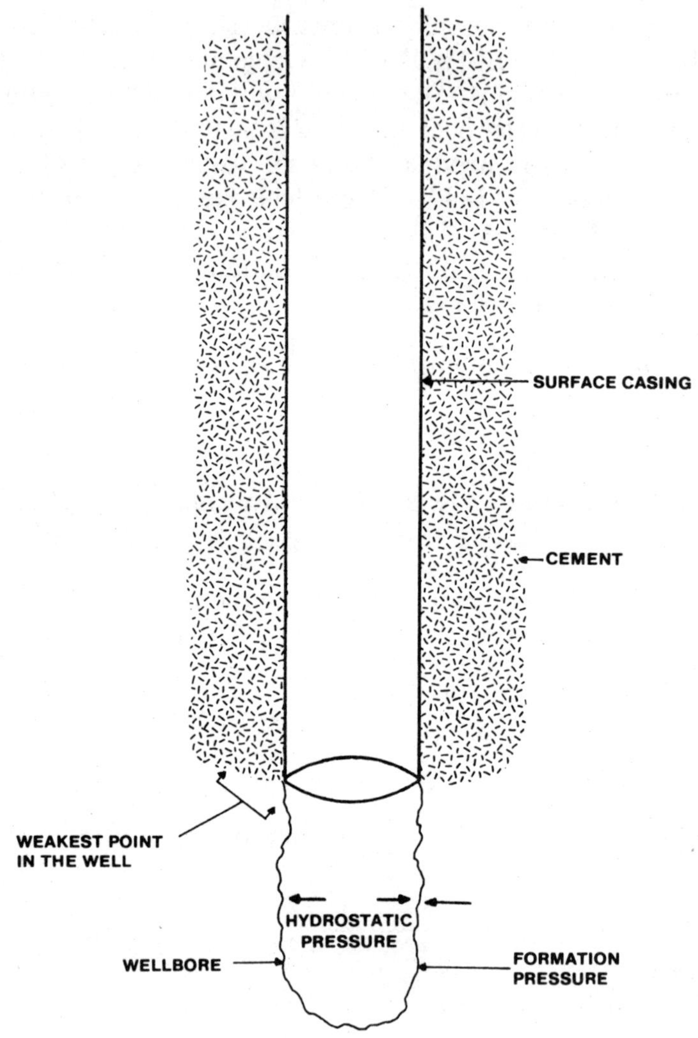

Figure 9-2. Testing the shoe.

annular preventer, using the rig or a cement truck pump. Hold the 721 psi for five or ten minutes. If it holds, bleed the pressure off and begin drilling. (Note: When pressuring up, do it only 100 psi at a time until 721 psi is reached.)

If the shoe does not hold the pressure, and it bleeds off, determine what pressure the shoe will hold to see if a squeeze job is needed. For example, if the pressure bleeds back to 300 psi, calculate the EMW to see what the shoe will hold. Take the existing 1,394 psi and add the 300 psi to it.

$$1{,}394 + 300 = 1{,}695 \text{ psi}$$

Then divide back to find the EMW

$$1{,}694 \text{ psi} \div 0.52 \div 3{,}014 \text{ ft (depth)}$$
$$= 10.8 \text{ ppg (EMW)}$$

Since the shoe will only hold a 10.8 ppg mud weight, a squeeze job will have to be run. Inform the operator of this development, and if your advice is requested, tell the operator to run a squeeze job to be safe. It would be a mistake to drill with a weak shoe. If you do not get the shoe to test and you hit high-pressure gas, the shoe can fail. Sometimes the breakdown will channel to the surface and crater the rig (see Figure 9-3). Testing the shoe is serious business. When the shoe is being tested you should be on location and watch the gauge personally. This will keep the hands from cheating on the test. In hard rock areas the shoe does not have to be tested because of the greater fracture pressure needed to break down the formation.

Call a service company that has a squeeze tool, and they will furnish a tool hand to run the tool. Then order out the cement and a truck to pump it. Normally on a surface squeeze 150 to 200 sacks of cement are needed. Some consultants prefer 200 sacks on the first squeeze, depending

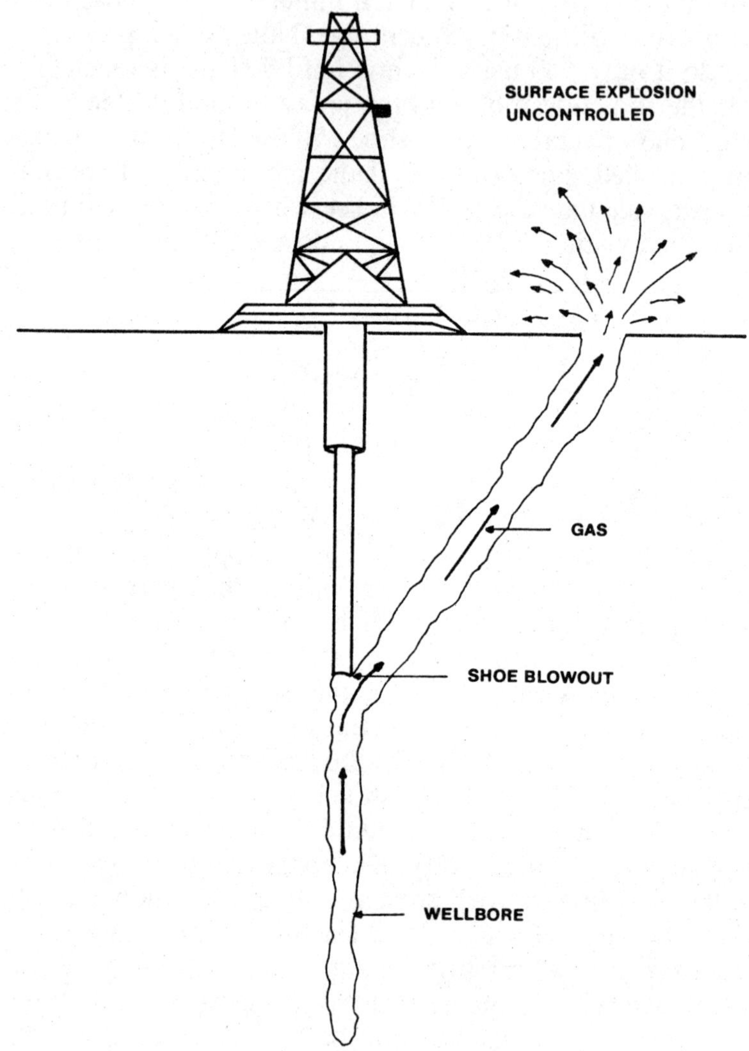

Figure 9-3. An underground blowout.

on how fast the formation takes the pressure. A decrease in pressure means mud is being pumped into the formation. The faster the pressure loss the more cement will be needed. Order out two loads of cement, so if the first 200 sacks do not squeeze off the formation, a second load will be ready to mix and pump.

Pull out of the hole while you are waiting for the squeeze tool hand and the cement. Decide where you need to place the squeeze tool so you can tell the tool hand where to set it. If, for example, you want it to set 15 bbl from the shoe, perform the following calculations to convert barrels into feet. First look in the cement book for the barrels per foot for a 10¾ OD 40.50 lb/ft casing. There are 10.19 ft/bbl. Multiply that figure by 15 (the number of barrels needed):

$$10.19 \text{ ft/bbl} \times 15 \text{ bbl} = 152.85 \text{ ft}$$

This means that you want the tool to set 152.85 ft (153 ft rounded off) from the shoe. Subtract this figure from the shoe depth of 3,004 ft.

$$3,004 - 153 = 2,851 \text{ ft}$$

Next figure the capacity of the drill pipe. Say you are using 4.5-in. 16.60-lb/ft drill pipe. The capacity chart in the cement book shows that this pipe will hold 0.01422 bbl/ft. Calculate:

$$2,851 \text{ ft} \times 0.01422 \text{ bbl/ft} = \begin{array}{l} 40.54 \text{ bbl (rounded off} \\ \text{to } 40.5 \text{ bbl to displace} \\ \text{the squeeze tool)} \end{array}$$

You must leave 5 bbl in the casing under the squeeze tool. To do this determine the displacement under the tool by multiplying the 10.19 ft/bbl by 10. This yields 101.90 ft

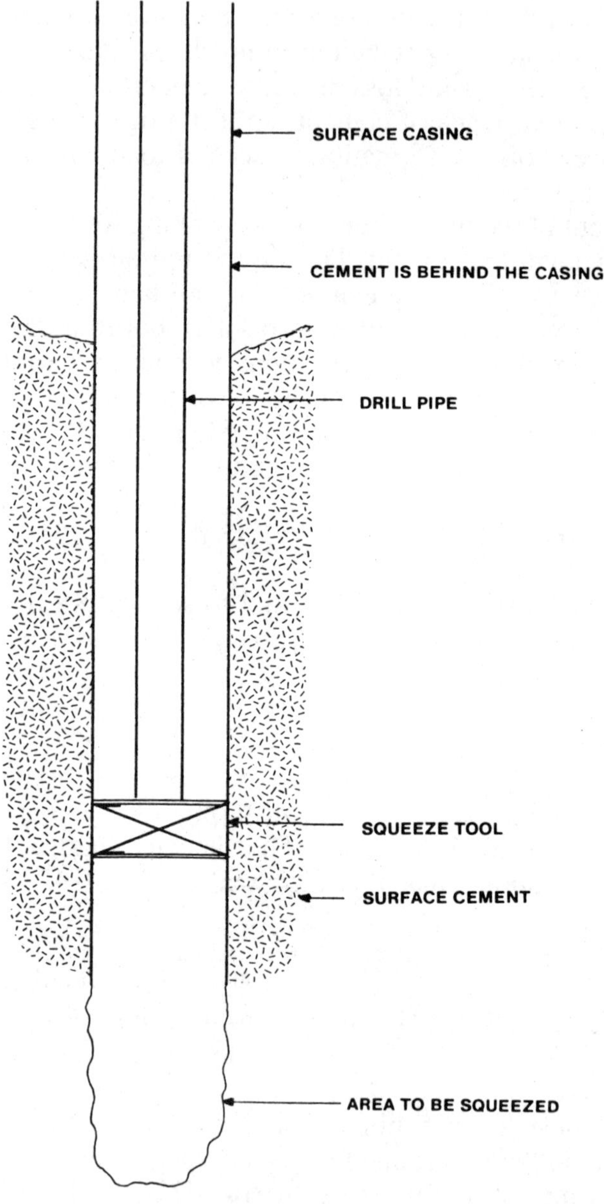

Figure 9-4. Positioning the squeeze tool for use.

(rounded off to 102 ft). From the vendor cement books, you will find that the casing capacity is 0.0981 bbl/ft:

$$102 \text{ ft} \times 0.0981 = 10.01 \text{ bbl (10 bbl rounded off)}$$

So the total amount of displacement fluid needed is:

$$40.5 \text{ bbl} + 10 \text{ bbl} = 50.5 \text{ bbl}$$

Next convert the sacks of cement to barrels by first finding the cubic feet:

$$ft^3 = \text{sacks} \times \text{yield (slurry yield)}$$
$$ft^3 = 200 \times 1.20 = 240$$

Then find the barrels:

$$bbl = ft^3 \times 0.1781 \text{ (constant)}$$
$$bbl = 240 \times 0.1781 = 42.74 \text{ (43 rounded off)}$$

Most consultants will spot the cement (that is, pump the cement to a designated place) by leaving the squeeze tool unset instead of bullheading the mud into the formation. Since we need to pump 43 barrels of cement, first pump 33 barrels, set the squeeze tool, and then pump the other 10 barrels. This will force the mud into the formation without causing any problems.

The cement was spotted because you don't ever want to pump cement around the tool. In this case, the pumping was stopped 10 barrels above the tool, and the squeeze tool closed, making it impossible for cement to be around the tool. This eliminated the possibility of getting the tool stuck or cemented in the hole—a very bad situation indeed.

Now pump the 50.5 bbl of displacement fluid to displace the cement into the formation and clear the tool to 51 ft above the guide shoe. Always add one barrel of displacement fluid to the calculation for fluid in the lines. So 51.5 bbl would be more accurate. (See Figure 9-5.)

If the cement pumps without any pressure build-up, there probably is not a squeeze. If, while pumping, there is a pressure build-up, raise the pressure from 2,500 psi to 3,000 psi and check for backflow. If there is no backflow, the shoe is squeezed. If it backflows, repump the cement back down the hole and check for pressure build-up. Hold for five minutes and then release the pump and check again for backflow. If there is still backflow, the shoe is not getting a squeeze; so try again. Sometimes after three or four tries the pressure will squeeze off. If it does not, close the cement manifold on the floor and hold that pressure for eight hours, then release it (see Figure 9-6).

During the last 10 barrels of the pumping operation of the squeeze, slow the pumping rate down to 0.5 bbl/min. Then on the last four barrels slow to 0.25 bbl/min. This will let the cement slowly squeeze into place and will not let a sudden build-up of pressure catch the cement operator off guard.

There is some controversy about the proper method for squeeze jobs. Some people prefer to release the squeeze tool and pull five stands of pipe to keep from cementing the tool in the casing. This is the poorest way to handle the job, because the purpose of the squeeze tool is to put the cement in place and enable the consultant to hold pressure against the squeeze to improve cement hardening and keep it from coming up the hole. If a squeeze tool gets cemented in the casing, the tool operator and the consultant have not correctly calculated the displacement.

If you do get a squeeze, release the tool and trip out of the hole, lay down the tool, make up the bit, and go back to bottom, short two stands. Next let the cement set for 18

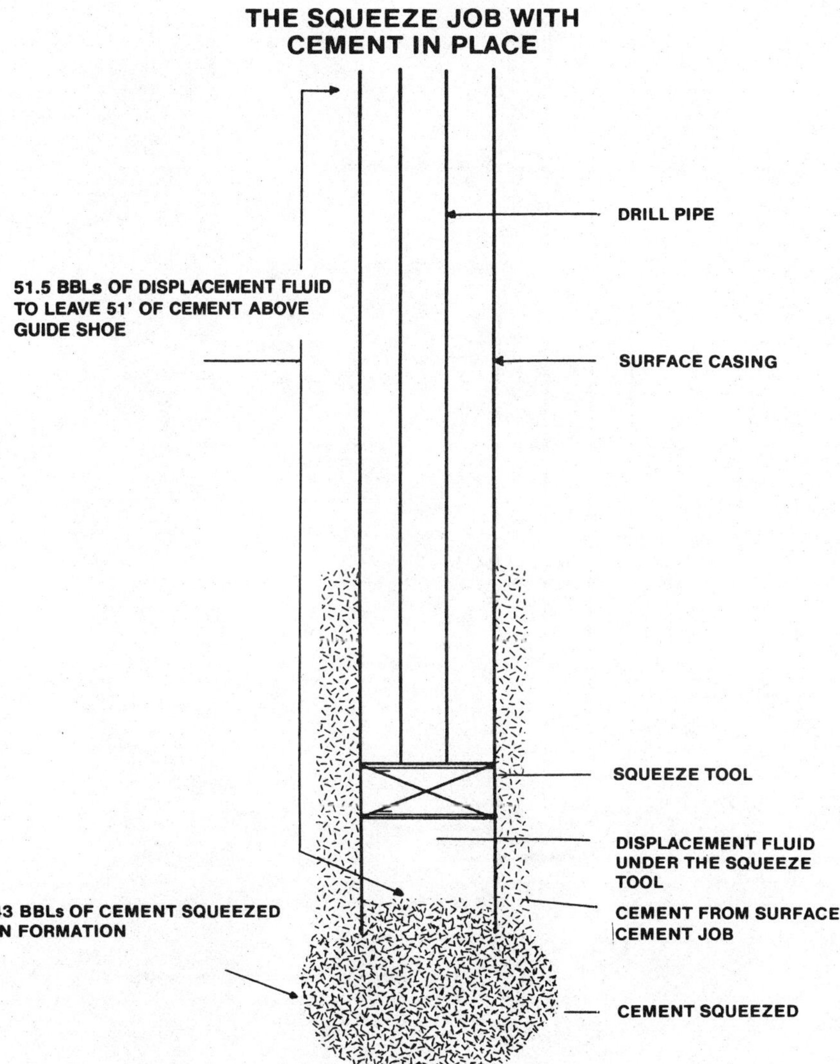

Figure 9-5. The squeeze job with the cement in place.

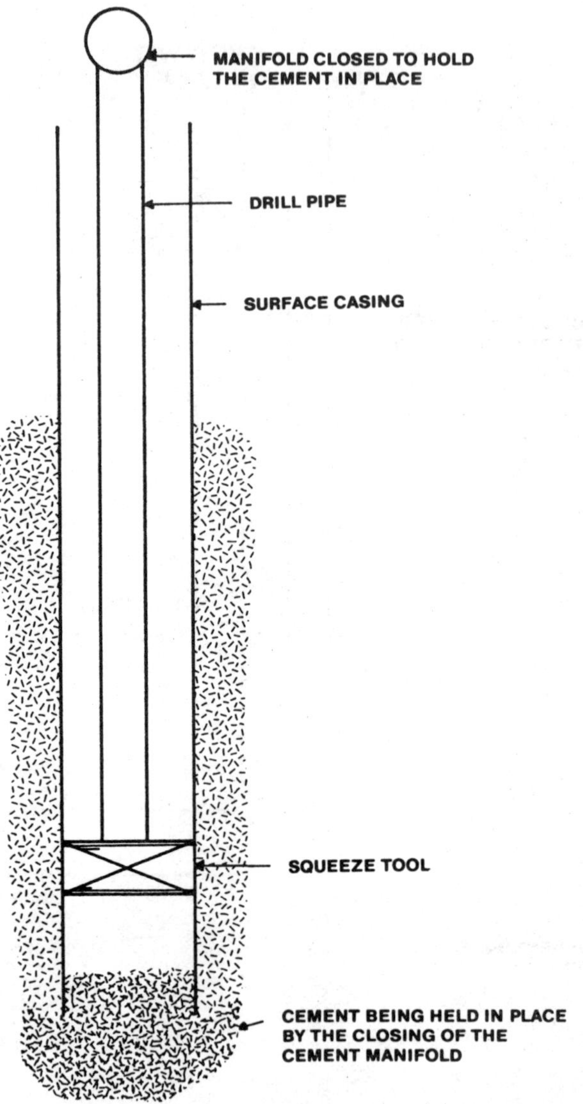

Figure 9-6. Closing the cement manifold to hold the cement in place.

hours, tag the cement, and record the depth of the tag. Then check the calculations for any errors.

As mentioned earlier, if there was not a pressure build-up when the displacement fluid was pumped, the squeeze was not successful. But that does not necessarily mean the formation will not hold. This is especially true in the Gulf Coast area. The gumbo and the soft formations will take the cement easier than anywhere in the world. It is like trying to cement sloppy mud. One of two things can be done:

1. Pump more cement down the hole in an effort to fill any hole or formation breakdown encountered.
2. Come out of the hole and let the cement set for 18 hours. Trip in and tag the cement. Then drill it out 6 ft below the guide shoe and retest it for 13.5 EMW (see Figure 9-7).

As you drill, give the cuttings time to come to the surface and check for green (not yet dry) cement. Drill 5 ft of cement, and circulate out the cuttings. If the cement is hard, it is okay. If it crumbles, shut down the pumps and let it sit six to eight hours more. Never drill out green cement. The vibration of the rotating string will cause the cement to channel and ruin the squeeze.

When cement squeezes, it actually squeezes the water into the permeable formation, leaving cement (solid particles) on the walls and making them stronger. Some Gulf Coast jobs require three or four formation squeezes before the show will test, so do not get disappointed if the first time does not work. Once you get a good squeeze, drilling operations can begin. Lay the squeeze tool down, set up your bottom hole assembly, and start drilling.

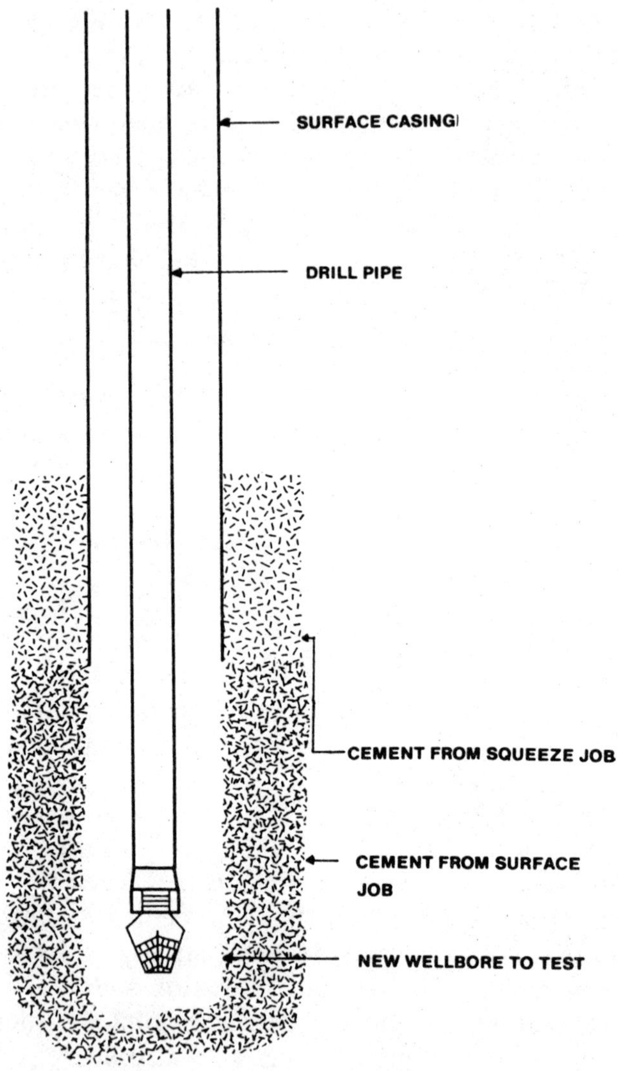

Figure 9-7. Retesting the squeezed formation.

10
Drill Bits

The selection of the correct drill bit for the well has always been a problem for the consultant. Making the right selection requires a careful study of bit records from other wells around the location. A successful consultant will sit down with the bit salesperson and go over bit records from no less than three wells in the area. Usually the geology is similar in the general area, but not always.

For the surface bit the choice is simple—a mill tooth bit. The bit salesperson will bring out the right one. For drilling out under the surface casing, a sealed bearing milled tooth bit is used. Formations drilled with the milled tooth bit are:

- Soft formation—Chalk, red bed, anhydrite, hard shale, dolomitic lime, medium lime rock.
- Medium formation—Hard anhydrite, hard shale, dolomitic lime rock.
- Hard formation—Hard shale with parite, sandstone, dolomite, limestone.

- Extremely hard formation—Abrasive lime, dolomitic sand rock, dense dolomite.

All of these formations can be drilled with milled tooth bits, with either sealed or journal bearings. The bit chart provided by your bit supplier will show which bits to use as the formation gets harder. The charts have been proven over the years and most are accurate.

When a well reaches the hard formations, the bit should be changed to a tungsten carbide bit. Some of the formations that require a carbide bit are:

- Soft formation—Lime, shale, unconsolidated sands.
- Medium formation—Dolomite, hard limestone, hard silicious shale.
- Hard formation—Sand rock, hard sandy limestone.
- Extremely hard formation—Taconite, granite, quartzite.

A tungsten carbide bit is normally called a ''button bit'' and has a very long life of 90 to 120 hours. On the Gulf Coast button bits are used mainly on deeper holes. Most wells to 10,000 ft use only milled tooth bits, but that depends on the geology. The life of a bit depends on the weight run on it, the rpm, and the geology. A sealed bearing milled tooth bit will last 18 to 22 hours, based on 15,000 to 35,000 lb of weight and 70 to 160 rpm. A journal bearing mill tooth bit will last 25 to 40 hours, at 20,000 to 30,000 lb and 90 to 110 rpm. On shallow wells along the Gulf Coast, this bit is used more than any other.

A journal bearing tungsten carbide bit will last 90 to 120 hours, at 25,000 to 35,000 lb and 50 to 70 rpm. It is best run at 60 to 64 rpm for good service and is used on most deeper holes on the Gulf Coast and hard rock areas. It is desirable because of its long life.

When drilling sand and shale at shallow depths, the bit works best around 90 to 120 rpm, but this cuts short the hours it will last. Check it at 70 hours, remembering rpm and weight determine the life of a bit.

The jets on a bit control hydraulics. The right combination of jets will make hole faster and more efficiently. Sizes 10/32 and 13/32 are the most widely used in the field, depending on the size of the slush pumps. The toolpusher will be of great assistance since he knows his rig and can advise you on what has been successful at each depth in the past. Many companies offer a drill package for calculating hydraulics on a hand held calculator or computer. Good hydraulics will drill the hole faster and increase bit life, and these calculations help to make drilling a science.

Since bits are such a small item they are easy to steal. Always order just what is needed and store them in a bit box or next to the door of your trailer. Only you and the toolpusher should have a key to the bit box. The best deterrent to theft is to watch the bits closely.

The diamond bit is used in hard rock areas and in deeper holes all over the world but rarely in shallow holes. It is designed for long life to reduce tripping. Cost is an important factor since diamond bits are much more expensive than tungsten bits. Diamond bits are invaluable in deep holes and in directional drilling. In hard rock areas they save time because they stay on bottom longer and are especially useful if there are kicks. Care must always be exercised when tripping in high pressure and unstable zones. A diamond bit will save a lot of trouble in these areas. Another point in favor of the diamond bit is its reclaim value. When returned, the diamonds are measured and the diamonds are bought back in the form of credit due the operator.

A diamond bit should be run with a little weight. The

weight is increased 2,000 lb at a time until the rate of penetration does not show an increase, then it is adjusted back to the penetration rate. The speed should be relatively high, about 110 rpm. If you use a downhole motor, the speed can really go up but use less weight. When making up joints and returning to bottom, the bit should be circulated off bottom for about five minutes to clean the bottom and increase bit life. Stability in the hole is very important, and a shock sub will keep the bit from bouncing on the bottom.

(H) In horizontal wells it is best to use a diamond bit, called a PDC bit. A good PDC bit can be used for up to five wells, so the cost is very low compared to that for rock bits and you never have to pull the bit until total depth is met.

A standard scale of 1 through 8 is used to grade the wear on bit teeth and bearings. A new bit is graded 0. Teeth are graded ''T''–(1–8) and bearings ''B''–(1–8). For example, teeth seven-eighths worn out would be graded ''T''–7. A worn-out bearing might be graded ''B''–4 meaning it is about one-half worn out.

Bit gauge is also graded on a 1–8 scale and marked ''G''–(1–8). Grading a bit takes experience and field grading is usually done with the unaided eye.

A point to note: If the bit is out of gauge, the next bit down will need to ream out 2 or 3 joints so the hole can be widened to prevent sticking. All bit conditions need to be reported so that future wells in the area can be drilled more efficiently.

The best time to pull the bit is determined by several considerations including:

1. The hours on the bit
2. The penetration rate
3. The footage drilled

4. The rpm and weight on bit (WOB)
5. If the bit is starting to torque up

If say, you have a 34- or 40-hour bit, hours on the bit are easy to determine. Get your bit records on the type of bit you are using. The usable hours will vary about two to four hours. For example if the bit you are using went 34 hours on the last run or has a 34-hour record on the bit report for the current depth, then start watching the bit at 30 hours. The hours on the bit are not the only thing to watch but this factor helps in determining when to pull the bit.

The penetration rate is also important to watch. If the bit slows down at 20 hours and you expected 34 hours from the bit, do not worry. You are probably in a hard formation, and the bit is cutting slower. Watch for an increase in rate of penetration. If there is none, then start watching the bit closely earlier, such as at 29 hours, since you are in a harder cutting area. If the bit slows down and you are at 30 hours, do not take a chance on dropping a cone. (The consequences of doing that are discussed later in this chapter.) Pull it since the time frame is correct anyway.

The footage is important because each bit is expected to cut so many feet, but this alone should not influence your judgment, since so many other factors are involved. For example, if your last bit went 545 ft and took 34 hours and you now have gone 551 ft in 31 hours, you are in the right footage range to pull the bit, compared to the previous bit.

The rpm and the weight on bit have a lot to do with the hours you can run the bit, because the more weight or rpm there is the more you wear the bearings on the bit. For example, if in your last bit run you used 90 rpm and 30,000 WOB and you went 34 hours, and on the next run you put 25,000 WOB at 85 rpm, the bit will last more hours. Still, after 34 hours you should watch it closely. Always remember that the WOB and rpm are directly correlated to the hours run, and you should compare bit records often.

If the bit starts torquing up with 20 hours on it, the problem may not be with the bit but perhaps with a keyseat, shale problems, etc. If, however, you have 29 hours on the bit and it starts torquing, check it carefully to make sure it is the bit that is torquing. Shut off the rotary table, then kick it in and check for torquing. If the bit torques pull it—do not take a chance at losing a cone.

If you consider all five previously stated factors together instead of singly, you will have a better idea of when to pull the bit.

Most bits have an estimated life range and will run well for that period. However, sometimes the bit will not torque up until the last moment—then it is too late, and you leave a cone in the hole. Leaving a cone in the hole is serious, since it requires a fishing job. And anytime you have a fish, you are probably in for a good deal of trouble. If you have to fish out a cone, use a globe basket or a magnet. Sometimes if you cannot get the cone with a magnet or too much cone trash is in the hole, you can use a bit mill to grind up the cone.

On the subject of used bits, I have used many on horizontal wells, since they are so economical. You need to find a good used-bit company that buys only good used bits from offshore drilling rigs. Offshore drilling rigs never put a used bit back in the hole. Some bits have been used for only 4 or 5 hours. I also stay with the top brands. Never buy a used bit from the less popular companies. They are less popular because their bits do not necessarily run the same each run.

Most consultants watch the hours on the bit and the penetration rate, and when it is time, pull the bit. No one can condemn a consultant for pulling a bit that has enough hours on it even if it comes out a little green. It is important to ensure that the drillers apply the proper weight and rpm to get the full life from the bit. You must watch the hours and the torque gauge, since it is your decision to pull the bit. With the help of the toolpusher and drillers there should be no trouble pulling the bit on time and avoiding a fishing job.

11
The Mud Program

The mud program is an important part of drilling a well. Although this chapter covers the basics, it would be wise to read several books on drilling muds to acquire a working knowledge of the subject. It is important for the consultant to be knowledgeable about mud engineering.

The mud circulates through the drilling rig as follows: First the mud comes out of the mud tank and goes to the mud pump. Next it goes through the standpipe to the swivel, down the kelly and the drill pipe, and out the drill bit. Then it goes up the annulus to the flow line and to the shale shaker. At the shale shaker the cuttings are separated and the mud goes back into the tank. This route is then repeated. (See Figure 11-1.)

1. Cool the bit and drill string for longer bearing life and less pipe damage from heat.
2. Bring cuttings to the surface. A good mud will keep cuttings from sticking above the bits or collars. A clean hole is very important.

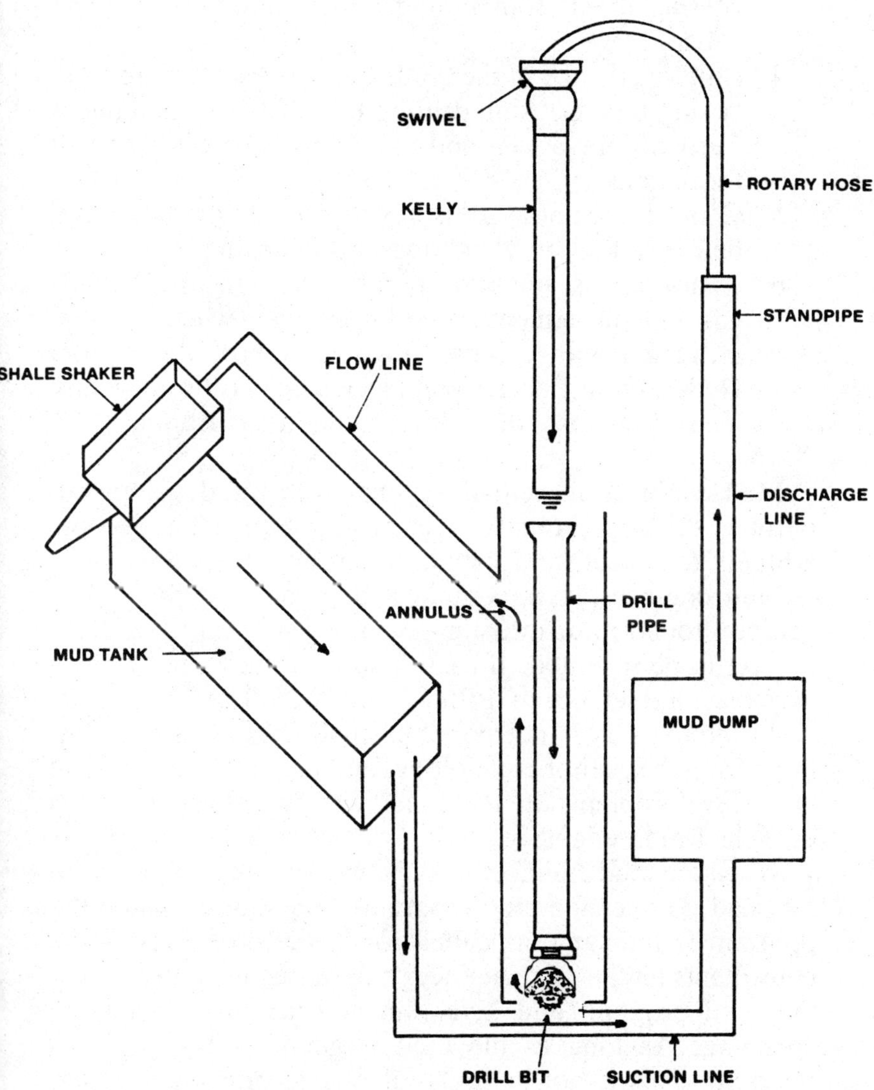

Figure 11-1. The mud circulating system.

3. Suspend the cuttings when the pump is shut off preventing them from falling down around the bit and collars.
4. Build wall cake in the bore. Cake keeps the bore from caving in and losing drilling fluid. Mud cake reduces formation invasion, and improves electric logging of formations.
5. Control downhole pressures (downhole pressures can be controlled by weighting-up the drilling fluid).
6. Guard against the invasion of hydrogen sulfide (H_2S) gas and subsequent corrosion of pipe (when the mud is treated with certain corrosive control chemicals). Hydrogen sulfide gas will make metal brittle and cause a string to separate, which results in a fishing job.

Mud engineering actually started with the development of rotary drilling. From its experimental starts, it has become a highly technical field with constant improvements.

Mud engineering is complicated because of the varied geology found in different areas. The perfect drilling fluid has yet to be invented. If each hole were the same with no variables, the science of drilling fluids would be more exact. Even though this is not possible, mud engineering has attained a high level of technology, and the consultant should have few problems in the field with the assistance of a confident mud engineer.

When the mud engineer arrives, you should sit down with him and go over the mud program. You should ensure the program is the same as called for in the prognosis. Most consultants let the mud engineer have a free hand in running the mud program and give him considerable latitude in spending. As long as the mud is good, you should not interfere. This is not to say you should completely ignore the mud engineer's work; but a good engineer will do his best, and he should be allowed to do so unencumbered. Still, it is your job to supervise, so not one sack of chemicals should go unobserved. Make sure that any torn or damaged sacks are used in the mud to further reduce the cost. Many mud engineers will not use a torn sack of chemical unless so instructed, but the operator will pay for it regardless. So keep your eyes peeled for torn or damaged sacks.

The derrickman is the person on the crew who takes orders from the mud engineer. The derrickman is actually the arms and legs of the mud engineer. So experienced derrick hands, in some cases, know as much or more than the mud engineer. If the mud engineer is not very good, your derrickman will quickly let you know. Get to know the derrickman on location so you can get answers when you need them. If a mud engineer cannot handle the job, order a replacement. *Get the service you pay for.*

There are hundreds of mud companies in the business today, but they offer basically the same service and the same product under different names. The basic additives to the mud include: barite, gel, fluid loss chemical, and lost circulation material. If the consultant understands these products, he can drill anywhere in the world.

Barite is used to weight-up the mud and control abnormal pressures. (The available supply for world use in the drilling business is limited. Eventually a replacement will have to be found.) Gel is used to build up viscosity and build cake in the wellbore. Adding gel to the system keeps the cuttings suspended in the fluid and moves them to the surface. Fluid loss chemicals reduce fluid loss in the wellbore and help control well cake thickness.

Lost circulation material (LCM) is used to fill areas where the formation is taking fluid. LCM is made of many materials; including fine, medium, and coarse graded mica; wood materials; cellophane flakes; fine, medium, and coarse nut shells; tree bark; cotton seed hulls; and ground plastic. Drilling in lost circulation zones requires the mud engineer and consultant to be on their toes. Lost circulation could result in a blowout, the sticking of drill pipe, possible formation damage to a pay zone, lost drilling time, and a large mud bill due to such zones.

Unless the well is a wildcat (a well drilled where no one else has drilled), the mud report will tell you where most lost circulation zones can be expected. Lost circulation material is sometimes necessary throughout the drilling process and is circulated in the mud system to stop the problem before it begins. The accompanying wear on the mud pumps will usually be indicated by a washout during drilling operations. It is very necessary to keep standby pumps in good order in a lost circulation area so the main pump can be repaired if it washes out. Lost circulation material should always be kept on location since lost circulation could occur at any time.

The following muds are used in the field:

1. Fresh water muds
2. Salt water muds
3. Oil base muds
4. Surfactant muds
5. Emulsion muds

The operator will tell you on the prognosis what mud is going to be used. Each one has a different use in different areas. The mud company is familiar with what muds have been more successful, and they will have the program already set up.

To discuss all the variations in mud would require a course in mud technology. In the field the consultant must rely on the mud engineer. He can explain why each mud is used and the specific characteristics of each mud used. Each day the mud engineer will submit a report on the mud and the results of his tests. Through these tests the evaluation of what chemicals to add to keep the mud correlated with the depth being drilled and the mud program will come.

In drilling, the ideal mud flow is turbulent as opposed to laminar. Turbulent flow allows the cuttings to come to the surface flat rather than tumbling. Laminar flow tumbles the cuttings and decreases the annular velocity. Keeping the cuttings in turbulant flow will also aid in cleaning the wellbore (see Figure 11-2).

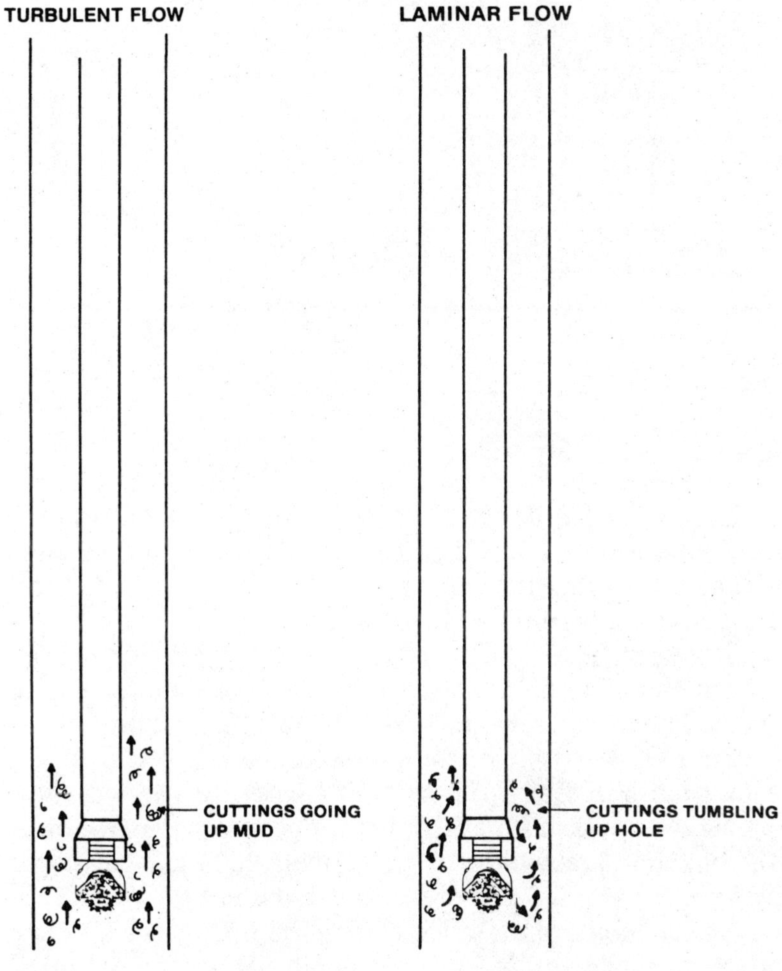

Figure 11-2. Different types of mud flows.

12
Drilling Ahead

Drilling ahead (when drilling is going smoothly) is the best time for the consultant to relax and catch up on some sleep. Up to this time things have been moving at a fast pace. With a bit on bottom and 30 to 40 hours to drill, you will have time to look over the rig and check for needed repairs or safety problems. It will also be a time for you to get to know the crew and watch them in operation.

On a 10,000 ft hole, for example, the last 4,000 ft will take from 6 to 8 bits (on the Gulf Coast), so you will have about 240 hours to square away the operation, train the crew on blowout prevention, explain their stations, get paper work in order, and prepare to finish the hole.

Always notify the service companies in advance and line them up while drilling to total depth (TD). This involves acquiring the following services:

1. The wireline logs
2. The casers

3. The cementers
4. The guide shoe, float collar, and centralizers
5. Thread cleaners for casing (a good service as crews clean and check the threads on the casing before pipe is run)
6. Locating casing and shipping it or having it ready to ship
7. The hydraulic wrench to nipple down

It is better to promise work four to five days before the job is finished allowing time to hear all the salespeople deliver their pitches and to compare prices. Some companies and small operators are cheaper and can save the company money. Saving money always makes the consultant look good to the operator.

It is your job to run a safe and reasonable operation. Many product and service companies are living on past reputation. Do not automatically call the big name company. In some cases smaller companies may give better service and may have hired the best hands from the larger operations. Most consultants will work for small independents that depend on saving money. At the end of the job this can add up to big savings to small producers. By looking for savings your services will be free to the producer when the job is finished.

Assuming there are no problems while drilling, everything should be ready to go by the last two days, and if planned correctly, the operation will be smooth right to the end.

(H) Now comes the horizontal well. Much has to be done after the surface casing is set and you are drilling to the vertical TD. Before the well becomes horizontal, the consultant must rig up the PWD equipment and line up the rental drill pipe, the handling tools, and the rotating head. A good consultant starts safety meetings on all phases of

Figure 12-1. A good kick. (Courtesy of Davenport Horizontal Drilling Consultants, San Antonio, TX)

the kick (Figure 12-1), especially if this is the crew's first horizontal well. The war with nature is about to begin. Horizontal wells tax a single consultant, and when it is over, he is totally exhausted from lack of sleep and having to make safety decisions quickly. Two consultants on the horizontal portion are what I recommend.

13

Produce-While-Drilling (PWD) Equipment

When the first horizontal well was drilled in the Pearsall field, PWD equipment did not exist. Through trial and error the system came into existence. The primitive system was something to see. Yet through the efforts of many engineers and consultants, the PWD equipment became a reality. H. S. Sadler of Houston, Texas, pioneered much of the equipment used today. Duke Davenport, "Doc" Holliday, Charlie Winn, and Buck Delano were also pioneers in the creation of better systems to make the business safer. Then a large service company named Sweco Oilfield Services, a division of Environmental Procedures Inc. of Houston, Texas, amassed all the equipment under one roof and rented out a complete system. As of this writing, it is still the only service company that furnishes all the PWD equipment in one package. Sweco will spot the equipment, nipple it up and nipple it down, and move it with the rig. The equipment contains the following:

1. Superchoke with control panel (Figure 13-1), used to regulate the gas and oil flow while a kick is taken.

Figure 13-1. Using the superchoke to flare gas. (Courtesy of Davenport Horizontal Drilling Consultants, San Antonio, TX)

2. Flow line to the gas buster (Figure 13-2). Once you drill horizontal, you bypass the shale shaker and send the returns through the choke and flow line to the gas buster and into the separation tanks or skimmer tanks (Figure 13-3).

3. The gas buster is used to separate the gas from the returns, and it sends the gas to the burn pit by means of a flare line. Then it sends the oil, cuttings, and drilling fluid to the first separation tank.

4. The separation of skimmer tanks allows the cuttings to settle to the bottom and the oil to skim to the next tank for treatment with chemicals. The separation tank can be cleaned out by opening a valve at the bottom (Figure 13-4). This cleans out the cuttings to make room for more fluid.

Figure 13-2. Flow line to gas buster. (Courtesy of Sweco Oilfield Services, a division of Environmental Procedures Inc., Houston, TX)

Figure 13-3. Separation tanks. (Courtesy Sweco Oilfield Services, a division of Environmental Procedures Inc., Houston, TX)

Figure 13-4. Cleaning out cuttings from a separation tank. (Courtesy of Davenport Horizontal Drilling Consultants, San Antonio, TX)

5. The generator for furnishing power to the separation tank pumps and the lighting. A smart consultant always runs another line to the rig's generator room to furnish emergency electricity in case the generator fails.

6. Pumps to transfer oil to the frac tanks for selling and to transfer drilling fluid back to the shale shaker to reuse it in the well.

7. Lights to keep the area well lighted at night. It is very important to keep the area lighted when a kick is taken. Also good lighting promotes safety around the rig, preventing one from stepping in holes or on tools, rattlesnakes, and flow lines.

8. Flare line with igniter. It is important, when gas is being produced, to burn it off at the flare line. An igniter sends a spark every 10 seconds, and any gas coming out will ignite.
9. Hoses and electric hookups. This is very important. By renting a package deal, the consultant's worries are over.

In some of the more exotic systems, there are shale shakers on the separation tanks, and the fluid is cleaned after it comes from the gas buster. This avoids having to clean out cuttings in the tank, and the oil and water going to the next tank are cleaner. Basically an operator can spend as much as he wants on PWD equipment. But remember, keep it simple to use because you have to teach it to field hands, not engineers. Another thing to remember is that field hands have not had formal training in any blowout schools, so they do not understand what is happening down the hole. Once a kick occurs, if any mistakes are made on the PWD equipment, oil can go all over the place. But if an emergency occurs, you can shut in the well with no problems, assuming that your blowout equipment is working properly. Since chalk horizontal wells are not high-pressure, the shut-in pressures will be around 300 to 900 psi. Sometimes the pressures are higher in certain areas, but this is rare. In the future new technology will enable drilling into other zones that will be high-pressure. The future of horizontal drilling looks good, because an operator can drill into several pay zones.

In summary, the PWD equipment (Figure 13-5) allows the operator to flow the well, send the oil to the frac tanks, flare the gas (Figure 13-6), and return the drilling fluid to the mud tanks. It is much like drilling a normal vertical well, but it produces oil (Figure 13-7) and gas. The oil, of course, can be sold.

Figure 13-5. Separation tank with cleaning equipment. (Courtesy of Sweco Oilfield Services, a division of Environmental Procedures Inc., Houston, TX)

Figure 13-6. A horizontal well flaring gas. (Courtesy of Davenport Horizontal Drilling Consultants, San Antonio, TX)

Figure 13-7. "Texas crude" after a kick. (Courtesy of Davenport Horizontal Drilling Consultants, San Antonio, TX)

14
Key Maintenance

While drilling the well there are certain key maintenance items to watch. If maintenance is performed properly, the problems of breakdown and lost time will be reduced.

The toolpusher is expected to keep the rig in good shape but this is not always done. Some drilling contractors have a bonus system for toolpushers who save money on maintenance. This is probably the worst program in the oilfield for saving money, since the rig rarely gets the repairs it needs, often resulting in major problems later.

The following is a list of areas to check for maintenance needs:

- Drilling line
- Crownamatic
- Pumps
- Blowout system
- Automatic driller
- Geolograph (recorder)

- BHA and redoping pipe
- Drill pipe
- Generators
- **(H)** Valve between shale shaker and rotating head
- **(H)** Roads
- **(H)** Rented drill pipe and handling tools
- **(H)** Separator tanks (skimmer tanks)
- **(H)** PWD pumps

Cut the drilling line. The drilling line is cut according to the ton/miles it has on it. Drilling contractors will cheat on this sometimes. This could result in the line breaking, and someone being killed or pipe being lost in the hole. If the driller has no record of the last cut, the drilling line should be cut on the next trip to be safe.

Check the crownamatic. The purpose of the crownamatic is to stop the block from traveling through the crown. This device has saved many lives and much property over the years. Each driller should test the crownamatic when he comes on shift. If it is faulty, it should be repaired immediately.

Check the pumps. Pump maintenance is important to the drilling operation. If the pumps are not maintained, problems can result downhole. Pump Number 2 must be in working order at all times so that if pump Number 1 fails, operations can continue until it is repaired. The best way to check pump Number 2 is to switch from pump Number 1 to Number 2 and drill for about one hour before switching back to Number 1. Some contractors may have a smaller pump as the Number 2. Although a smaller pump will slow down the rate of penetration, it is better to drill slower than not at all. It is very important to ensure proper maintenance of the pumps, so do not neglect this area.

Testing the blowout system. The drilling operation needs a pressure tested blowout system to guard against kicks.

Because of vibration on the rig the system needs to be checked often. The system should be retested every seven days to check for leaks around flanges, lines, valves, and the choke line and manifold. Repair all leaks and retest the system. If you do have a kick and a leak is present, the rig can catch on fire, causing much loss of equipment and sometimes a loss of life.

Test the automatic driller. If the automatic driller is out of adjustment, excessive weight on the bit and hole deviation or reduced weight on the bit and a reduced penetration rate can result. If the consultant feels the automatic driller is off, a service hand should be called.

Test the geolograph. The geolograph needs to be checked both before drilling is started and during drilling to ensure it continues to work properly. If it is not accurate, a service hand must be called to make repairs. The geolograph records footage per hour, string weight, down time, pump strokes per minute, etc., so it is very important to the rig operation.

Inspect the BHA and redope the drill pipe. Always inspect the BHA every three or four trips for washouts. Redoping helps to keep the threads sealed. It is best to redope drill pipe as often as possible. On triple rigs the drill pipe can be redoped every third joint with a different set of joints redoped each trip. If one or two joints are laid down, the stands will break at a different joint each time, thus all the tool joints will be redoped every third trip.

Check for bent drill pipe. If bent drill pipe is in the string it must be pulled, laid down, and replaced. You can spot bent pipe by looking at each stand as it is in rows in the derrick. Watch for bent pipe closely as such pipe could separate downhole.

Check the generators. The generators are very important to the operation. The driller should switch generators daily to make sure they both work. Also check the PWD generator, since it must work when a kick is taken. The following should be checked, too:

(H) *Test the valve between shale shaker and rotating head.* Make sure that this valve is pressure-tested for leaks before horizontal drilling is begun. Test it to 2,500 lb when the last test is done. This valve must work and must not leak. To be completely safe, two valves could be installed in series and both closed.

(H) *Inspect the rented pipe and handling tools.* Before it is shipped to the location, pay an inspector to inspect the pipe. Normally a spot check is sufficient and will cost about $200. It could prevent many problems. Also the handling tools are normally junk, so make sure that the slips work and all the inserts are tight. Have the pipe inspector look at them also and report to you. He will probably do it while he is inspecting the pipe.

(H) *Clean and inspect the separator tanks.* These need to be cleaned and the valves inspected visually, by putting a little water in the tank. These tanks do not hold anything but hydrostatic pressure so it's not critical, but they can be mess makers when a kick is occurring. A bag of gel poured around the valve inside the tank will normally seal it, to finish the job.

(H) *Test the PWD pumps.* The PWD pumps must be tested before horizontal drilling is begun. Often new packing will have to be installed. It takes only 5 to 10 minutes. The pumps carry oil to the frac tanks and water returns to the rig to be reused in drilling.

15
Special Problems During Vertical Drilling

Things do not always "turn to the right." And when problems occur, it is up to the drilling consultant to solve them. Some of the problems consultants encounter include:

- Lost circulation
- Controlling hole deviations
- Sticking and torquing pipe
- Equipment failure: bits, tools, rig, pumps
- Bridging
- Going back to bottom
- The twist-off
- Pipe washout
- Strapping pipe to get accurate tally

Lost Circulation Problems

Lost circulation, as discussed in Chapter 8, is a very expensive problem to deal with. It is usually caused by drilling into a permeable formation or faulted, fissured, or jointed zones. To cure most zones, lost circulation material (LCM) is added to the mud system. The mud engineer can prescribe what has been most successful in the area of the well and usually has this information on location.

When a drill bit penetrates a lost circulation zone, the usual procedure is to pull the bit one foot off the bottom and reduce the pump strokes to about one-half normal operation. This will reduce the equivalent circulating density (ECD) and allow time to mix the LCM into the mud. The rate at which LCM is mixed depends on the size of the pumps and the volume pumped. Lost circulation material must be mixed swiftly to solve the downhole problem. A quick calculation can be made to determine how fast the LCM will hit the lost circulation zone.

Assuming there is lost circulation at the bottom of the wellbore, the formula is:

surface-to-bit time
(S to B)(in minutes)

$$= \frac{(\text{bbl/ft DP})(\text{depth}) + (\text{bbl/ft DC})(\text{depth})}{(\text{bbl/stroke})(\text{strokes/min})}$$

DP = drill pipe

DC = drill casing

Example

At 8,000 ft with a 4.5 in. XO drill pipe weighing 16.6 lb/ft to 7,408 ft, the capacity can be found from the cement book to be 0.01422 bbl/ft with a BHA of 592 ft and a capacity of 0.01776 bbl/ft. Calculating 0.09 barrels per stroke and 60 strokes per minute, we get:

$$S \text{ to } B = \frac{(0.1422 \text{ bbl/ft})(7{,}408 \text{ ft}) + (0.1776 \text{ bbl/ft})(592 \text{ ft})}{(0.09 \text{ bbl/stroke})(60 \text{ strokes/min})}$$

$$S \text{ to } B = \frac{105.34 + 10.51}{5.4} = \frac{115.85}{5.4}$$

$$S \text{ to } B = 21.45 \text{ min}$$

As calculated, in 21.45 minutes the LCM will hit bottom. So after a short circulating time, the problem should be solved. If the problem is serious, the shale shaker can be bypassed to keep from losing the LCM. Bypassing the shale shaker permits the LCM to be repumped down the hole rather than lost in the screen of the shale shaker. Once the mud pits stop showing a loss, drilling may resume, but must be watched closely. (See Figure 15-1.)

It is also important when fighting lost circulation to keep the pipe moving, since the pipe could become stuck due to the permeability of the hole and the possible heaving of the formation. The pipe should be rotated for five minutes and then the kelly lifted and reamed back down. Lost circulation is dangerous and should not be taken lightly.

Many blowouts have resulted from mud going into the formation and gas coming back up the hole (see Figure 15-2). When liquid mud is available, the consultant should have it sent to location so he doesn't run out of mud. Always remember to reduce the pump to one-half normal operation range to give you more time to mix the LCM with the mud.

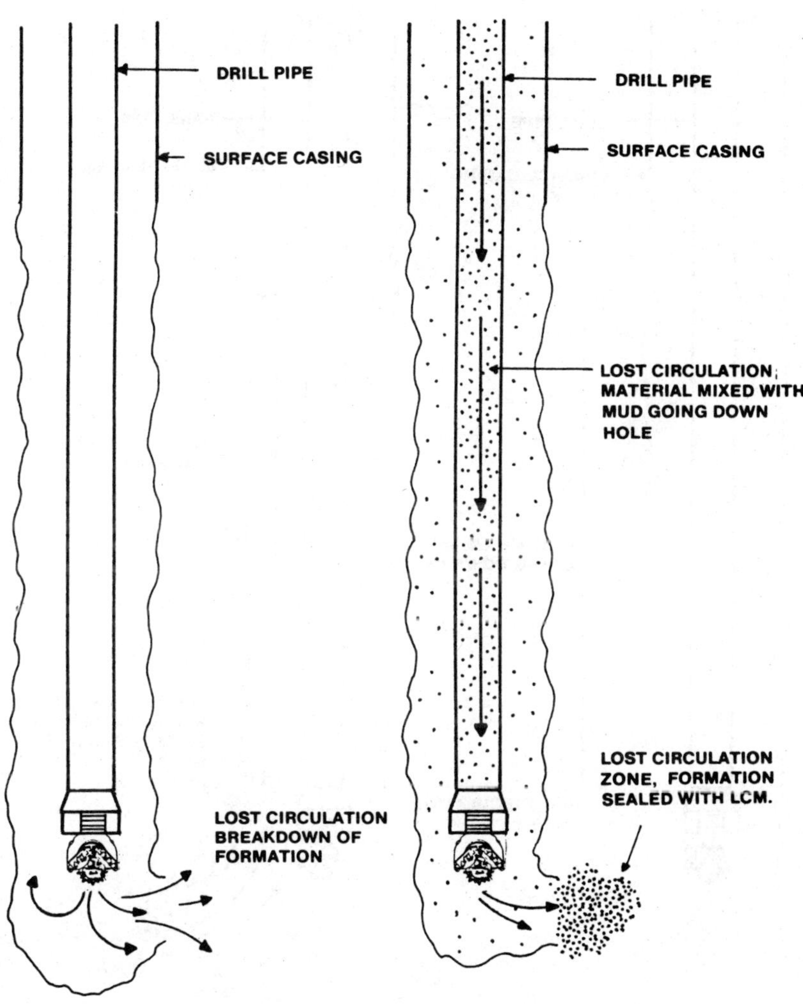

Figure 15-1. A lost circulation zone.

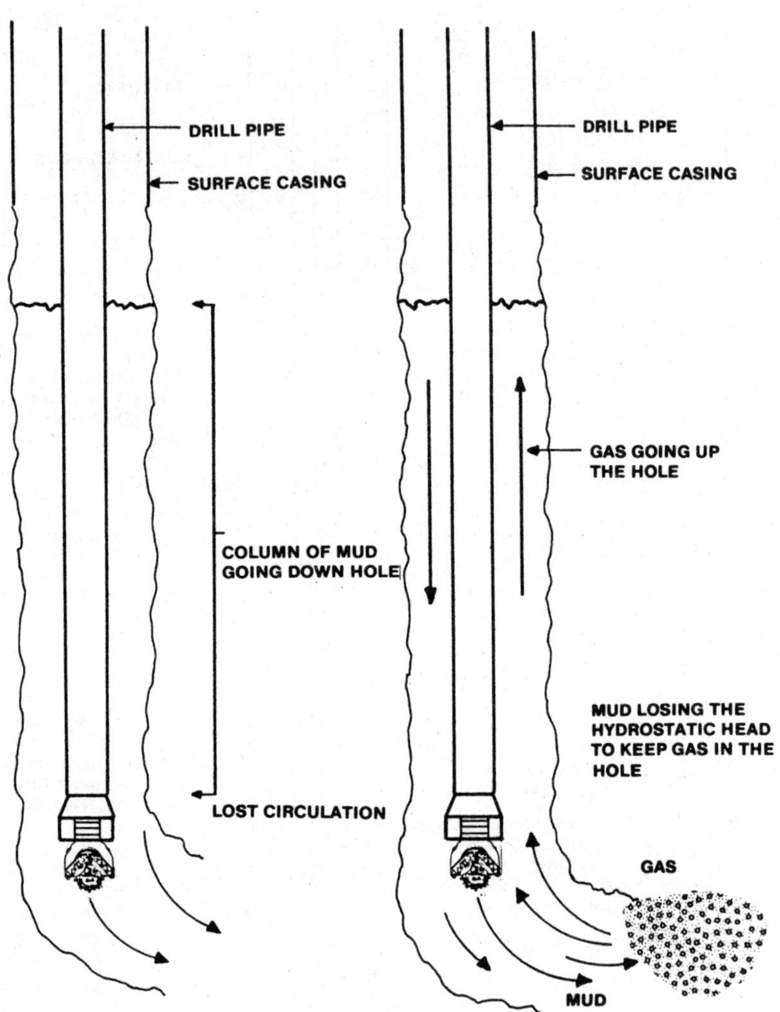

Figure 15-2. Underbalanced mud in a lost circulation zone.

Sometimes the lost circulation problem takes days to solve and can cost thousands of dollars. If the zone cannot be sealed, it will tax the brains of the consultant and the engineer. Putting a cement plug downhole sometimes will solve the problem. Then you can drill it out and go from there. Some wells have been abandoned because the zone could not be cured; however rare, this has happened. Sometimes pumping a large slurry of LCM downhole, pulling ten stands of drill pipe, shutting off the pumps, and letting the slurry sit overnight works.

When a lost circulation zone is encountered, the mud bill increases drastically. The operator may become alarmed but you need to stay calm during this problem so you can think clearly and keep operations running smoothly. Usually the problem can be solved.

Controlling Hole Deviation

Controlling hole deviation will not be a major problem if some simple rules are followed. First, you should take a survey every 500 to 700 ft while drilling. In the shallow part of the well a wireline survey can be run, but as the well gets deeper a "drop survey" will need to be run at the end of each bit run. (Most bit runs will be under 500 ft.) This will give a good picture of the hole while drilling. A deviation chart (see Figure 15-3) should be kept on the hole for a clear picture and for locating dog legs and keyseats. A chart will simplify following hole deviations.

The following is an example of what steps might be taken as the survey information is gathered.

Example:

For the first 6,000 ft the deviation is between 0.5° and 1° which is acceptable. Then the bit deviates 1.5°, and must be brought back on the right track. The first step is to

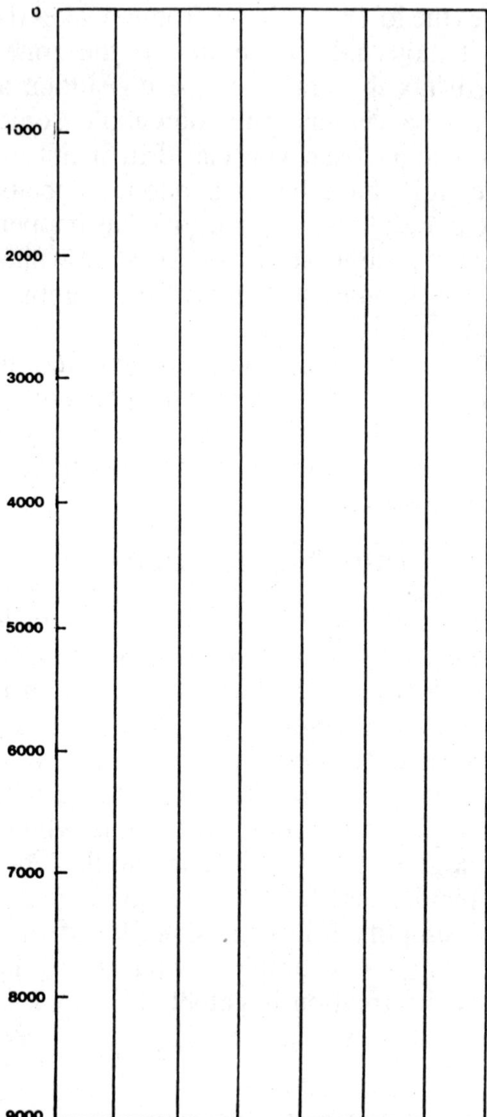

Figure 15-3. Hole deviation chart.

take weight off the bit by reducing the weight to about 25,000 lb. Pump strokes are increased eight strokes per minute, raising the pressure about 200 psi. This in turn increases the rpm about eight to ten turns—say, from 90 to 98 rpm.

The next survey at 6,800 ft shows a deviation of 2°—the deviation has slowed down but the hole is still going off course. The weight on the bit is decreased to 20,000 lb while the rpm is kept the same, in an attempt to use the weight of the collars to create a pendulum effect on the string and thus bring it back to center. The rate of penetration will decrease, but this approach is necessary. The time to stop the hole from kicking off is while you are drilling.

The next survey is at 7,300 ft and shows a 1.5° deviation. So the reduction of weight caused the string to bend back towards a straight hole. Weight on the bit should be increased to 25,000 lb and the rpm maintained about the same. As a result of the increased weight, the rate of penetration will improve. The next survey at 7,800 ft shows a 1° deviation, which means deviation is under control.

Next, a fault line is encountered, which in many areas will cause a rapid hole deviation, because the bit will follow the path of least resistance. The bit will follow the fault line as long as the fault goes undetected. The next survey at 8,250 ft shows 3° deviation, and immediate action is needed.

We have a good stable string, so we must rely on weight reduction again. We go back to 20,000 lb and the unavoidable decrease in penetration rate. The next survey is at 8,570 ft, and the deviation is 4°, so weight reduction has not done much to stop the bit walking in the faulted zone. The only thing to do is further reduce weight and start reaming each drilled joint down twice. This will widen the hole toward center (see Figure 15-4). The weight is reduced to 15,000 lb, with the resulting penetration rate being very slow.

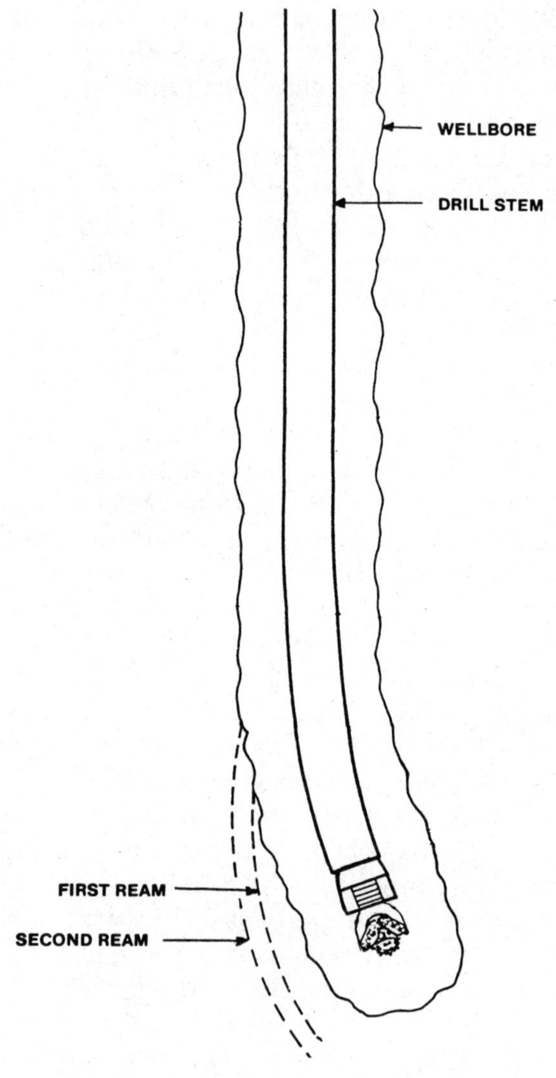

Figure 15-4. Reaming the hole to bring it back to center.

The next survey is at 9,100 ft and shows a 2.5° deviation. The hole is coming back in but a little too fast. A dog leg can be formed if the string comes back to center at too rapid a rate. The weight is increased to 20,000 lb to stabilize the deviation and to allow you to bring it in slower.

A rule of thumb to use to determine the footage to control deviation is 2.5 times BHA length. If you kick off, say 3° or 4° and the BHA is 575 ft, then 2.5 times 575 ft should be enough footage to keep a dog leg out of the hole. If you bring the string back slowly, stay calm and take your time, you will stay out of trouble.

When a well kicks off, it kicks off in a spiral path, not a vertical one. The hole, if it could be seen, looks like a corkscrew. This is why it can be brought back to center easily. By running a gyro or multishot survey, the hole can be accurately shown on a graph and the direction of the hole recorded.

Always keep the operator informed about what is going on and what is being done to correct the problems you have encountered. When in trouble keep the company engineer updated on successes and failures. Sometimes he may want to change and make decisions from his office, but this does not happen often unless he is new on the job. Once he makes a few decisions, he will generally allow the consultant to handle things from then on. If the operator's engineer radically changes your program, you should call your boss first and discuss what is going on. If your boss sees a problem, he should call the company engineer and discuss the idea with him before letting you use it. Most of the time when the consultant has reached that depth on the well, the operator trusts his judgment.

The main thing to remember in hole deviation is that reducing the weight and increasing the rpm will solve the problem in most cases.

Sticking and Torquing Pipe

Sticking pipe can be caused by a number of conditions including:

- *The hole sloughing in and around the bit or drill collars—* Most hole sloughing can be prevented by adding gel and by weighting-up some. If shale is sloughing in the hole, the mud can be treated with an asphalt-base chemical that will prevent water in the hole from getting behind the shale and pushing it into the wellbore or causing the shale to swell up and push out into the wellbore.
- *The mud not cleaning the hole properly—*In most cases gel will improve this problem by adding viscosity and bringing up the cuttings.
- *A dog leg—*This may cause dragging or torquing problems that could lead to the formation tearing up and sticking at the collars or around the bit. (See Figure 15-5.)
- *Keyseating—*This is caused by the pipe wearing into the side of the wellbore so that the string gets stuck as it is pulled through the keyseat. Normally a three- to six-point roller reamer will solve the problem by wiping the hole every trip. Also the stabilizer will help greatly in wiping the hole. (See Figure 15-6.)
- *Drill collars—*These can stick if there is a washout above the bit. If there is a washout, the pump pressure will decrease slightly until the washout gets bigger. The bit should be pulled off bottom and the system checked for leaks. If none are found, then the decrease in pump pressure is probably due to a washout. The consultant must always watch for this problem, because sometimes a new driller will kick up the pump strokes to maintain the pressure. That should never be allowed to happen since a twist-off could result or the string could become stuck below the washout because the hole is not cleaning properly due to less fluid reaching the bit. (See Figure 15-7.)

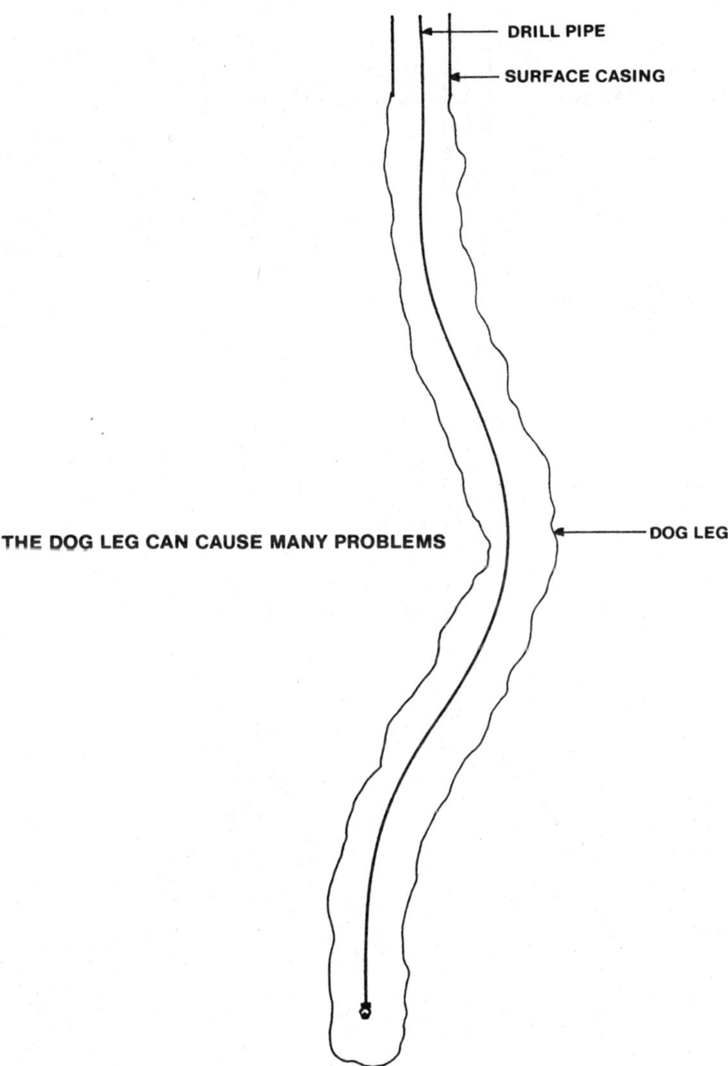

Figure 15-5. A dog leg.

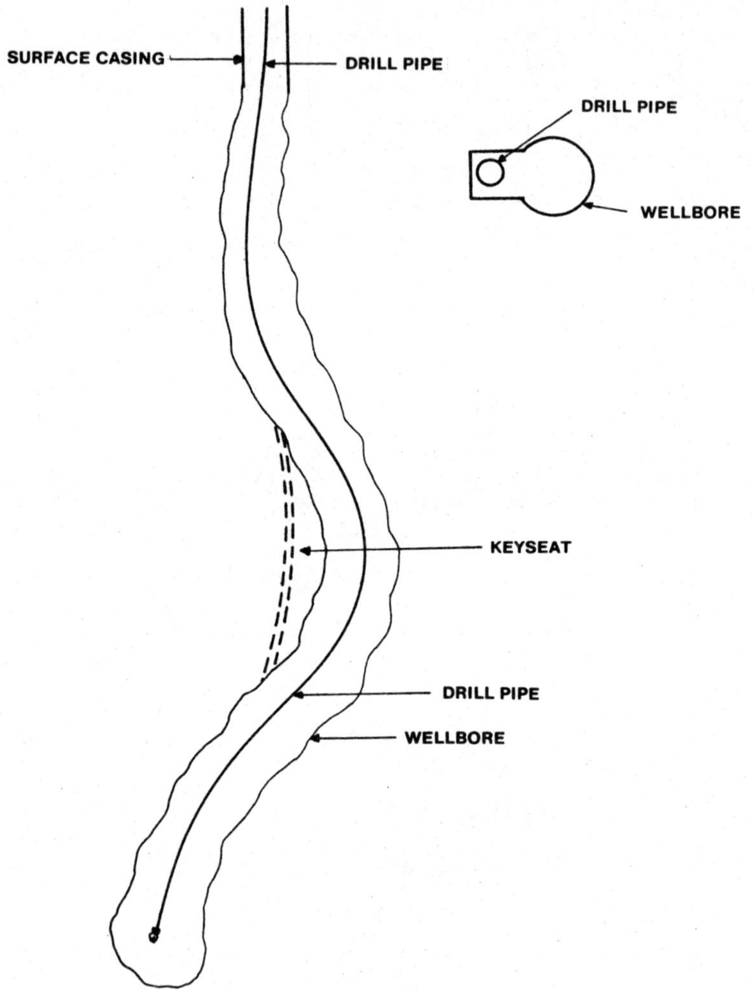

Figure 15-6. A keyseat.

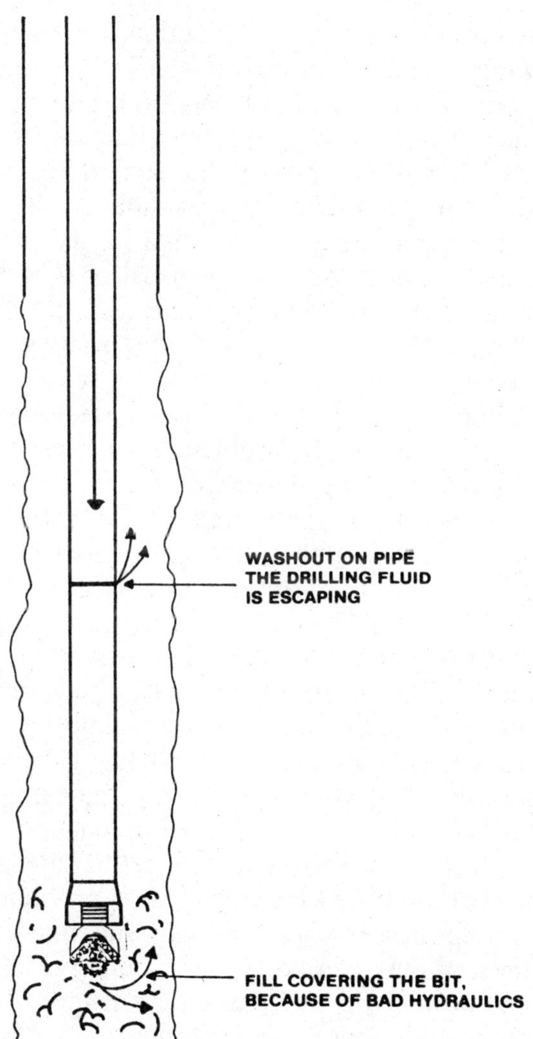

Figure 15-7. Problems caused by pipe washout.

Equipment Failure: Bits, Tools, Rig

Drilling bits are designed to function for a given period of time without failure; however, sometimes when the bit is placed on bottom it will not operate properly. The penetration rate may be slow, or torquing may occur after only a short time. It will be obvious that something is wrong.

To find the problem check all possible sources. For example, if the torque gauge shows that torque has been released, it may be that there is a keyseat or that the hole is not cleaning properly. If the pressure does not drop, it is not likely that a washout is causing the problem.

If the penetration rate slows considerably, then the bit is bad. It should be pulled and checked. Defective bits are rare, but they can be real troublemakers. Refuse to allow the operator to pay for a defective bit, since it has caused lost rig time, headaches, and great loss of money. Advise the bit company that if they charge for the bit, their bits will not be used in the future. As in any commercial business, including the oil business, a customer is entitled to a replacement or a refund for defective products.

Equipment failure is usually caused by poor maintenance or by using worn-out equipment that should have been retired. When the oil business is booming, it is harder to find good equipment and what is available may not be in the best shape. When the consultant gets on location, he should look for problems and ask the toolpusher to have them fixed before the well is drilled too deep. Most problems involve engines, pumps, drawworks, brakes, old lines, etc., and most of these things can be repaired quickly. However, if the rig goes down it is the contractor's responsibility.

Make sure the contractor's representative records all downtime. Most rigs are allowed a given amount of downtime per month, and an unrecorded hour here and there could cost the operator money. Some toolpushers try to cover downtime if the consultant is in town or asleep. The geolograph will give you the first indication that the rig was down, but some older drillers cover up the geolograph so the consultant must watch it closely.

Downhole tool failure is not difficult to prevent. All that is necessary is a record of the hours on the tool. By checking with the suppliers on the standard hours of operation, you can keep an acurate rotating hour chart on any tool in the hole, including drilling jars, shock tools, and stabilizers, in operating condition. Remind the pusher to check tools and replace them if necessary. Always check the elevators and slips. If their conditions are borderline, demand they be changed or repaired. It is better to wait on replacements than to have to account for downtime.

Bridging

Bridging, a common phenomenon, is the sloughing in of the hole due to some of the following:

- Improper viscosity
- Unbalanced mud
- Swabbing the hole when tripping out

In shale areas bridging is common, since water can get behind the shale and push it out into the wellbore or cause the shale to swell and fall into the wellbore (see Figure 15-8). (Bridging is what causes most fires and blowouts to put themselves out.)

If there is a reduction in weight on the weight indicator when tripping into the hole, there is a bridge. When that happens the string can be pulled back up through the area to provide a cleaning action. If the string becomes stuck, the kelly can be hooked up and the bit washed down to the bridge. When the bit hits the bridge, simply ream or drill

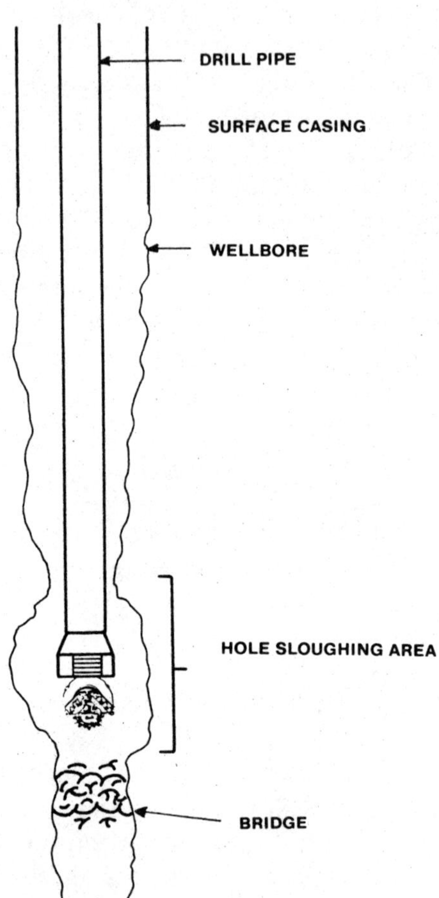

Figure 15-8. Bridging.

through it slowly. Then you can break the kelly and continue tripping in the hole.

Sometimes when you start reaming through a bridge a new hole starts and it becomes impossible to find the old hole. Occasionally a consultant has had to call in a report saying that the rig has lost 4,000 ft of hole. So when you ream through a bridge, go slowly and let the bit gradually wash its way down (see Figure 15-9).

Going Back to Bottom

When returning to the hole with a new bit, it is always a good idea to ream back to the bottom two or three joints to further widen and clean the bore. In shale areas doing this will give you some margin of safety against bridging or getting stuck. When tripping out of the hole, some fill will fall to the bottom, but reaming will keep fill from being a problem.

The Twist-Off

When drill pipe separates in the hole, it is called a twist-off. Check the geolograph to see if the driller is to blame. Most twist-offs are caused by one of the following:

- Washout in the drill pipe or in the drill collars
- Too much weight on the bit, which causes torque and parting of the pipe
- Encountering hydrogen sulfide gas. Hydrogen sulfide gas may come up in a fault from deep down even at shallow depths when it is not expected.

Once the problem is located, action must be taken quickly. The longer the pipe sits in the hole the more fill

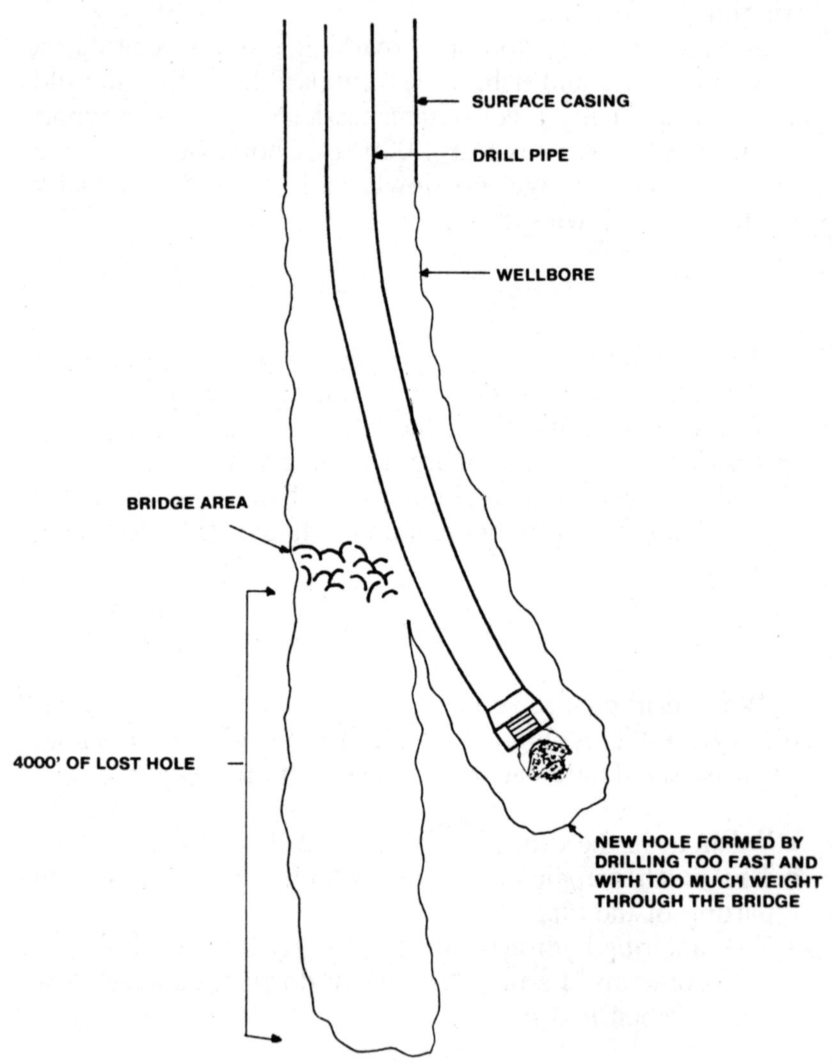

Figure 15-9. How a bridge can cause a hole to be lost.

will fall in around the pipe, which makes the pipe stick more. Weighing the string will help determine where the twist-off occurred.

Example

If the string and the block weighed 222,000 lb to begin with, but after the twist-off only 150,000 lb are indicated, the length of string in the hole can be calculated. First subtract the weight of the blocks (45,000 lb subtracted from 150,000 lb leaves 105,000 lb—the weight of the remaining string). (See Figure 15-10.)

If the drill pipe is 4.5-in. XO pipe at 16.60 lb/ft, divide:

$$\frac{105,000 \text{ lb}}{16.60 \text{ lb/ft}} = 6,325 \text{ ft of string}$$

If you were at a depth of 8,572 ft when the problem occurred, then the pipe parted at 6,325 ft, leaving 2,247 ft in the hole to be fished out. Remember that if you will draw a picture of what is in the hole when you have a problem, the problem will look easier to solve. In this particular case the blocks weigh 45,000 lb and the drill collars weigh 45,000 lb, so the string must weigh 132,000 lb. Write in the weights on your drawings so you will have an accurate picture of what is happening.

The next step is for the driller to trip out of the hole, while strapping the pipe. Then you call out a fisherman. Tell the fisherman that there is about 2,247 ft of pipe in the hole at the TD (total depth) of 8,572 ft. The fisherman will bring out the necessary tools.

Fishing is relatively simple, but is best left to the fisherman. Many consultants make the mistake of trying to fish the hole themselves. This is like a lawyer trying to defend himself. Fishermen are paid to fish and solve the problem.

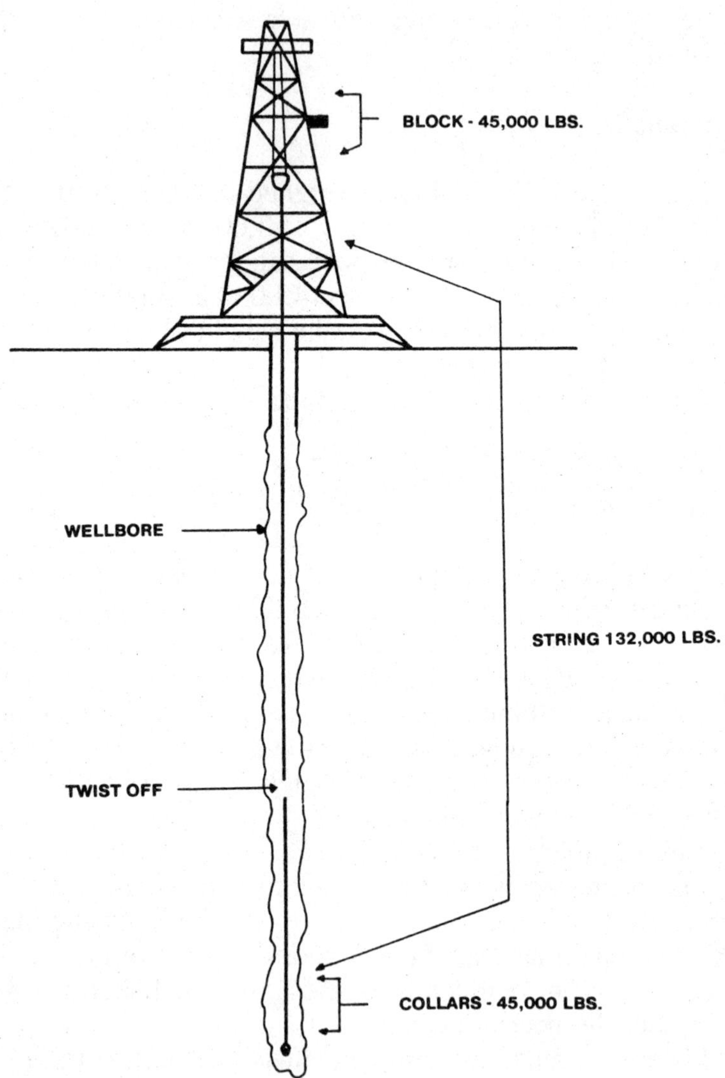

Figure 15-10.   A twist-off.

If you turn the job over to the fisherman, then his reputation is at stake and it relieves you of responsibility if something goes wrong. Let the fisherman explain his plan to you, so you can explain it to your boss and the operator. In most cases, by the time the string has been tripped out of the hole, the fisherman will be on location and ready to take over.

Pipe Washout

Pipe washout is a common problem because of the many times pipe is broken out and made up. Some of the causes of pipe washout are:

- Banging the pipe facing when making up the pipe
- Not using enough pipe dope

A driller should take enough time making up joints to ensure no damage occurs, since little nicks and pings can result in pipe washout downhole under pressure. Make a driller who is trying to set an in-and-out record slow down.

Pipe dope is designed to withstand heat and pressure—it has kept the oil industry functioning for years. The proper application of a good API dope will save many problems.

Redope drill pipe using the crossing method. In simple terms, this means breaking the stands at different joints each trip so that older connections are broken and redoped. The driller should lay down one or two joints at the beginning of the trip, then break the remainder of the string into normal stands. He should break the bottom hole assembly every third or fourth trip, check for washout in the facings, lay down bad drill collars, then redope. It is very important to keep the BHA in good shape. It is also important to check the XO (changeover) sub and the first joint of drill pipe, since they take a lot of torquing. Being cautious will keep the consultant on the job—and on the next one that comes up.

It is almost certain that every three or four runs a washout will be found in one or more drill collars and they will have to be laid down. Everyone will disagree with the order, but when the first collar is laid down, that will quiet people down and add to the oil company's confidence in the consultant. In a string of drill collars, one or two collars are bound to go bad every three or four bit runs. (You can count on it happening! In the past fifteen years out of twenty as a drilling engineer I have not lost one collar personally or while in charge of other personnel, because I inspect them faithfully and have my crew do the same.)

Hire an inspecting company to test the threads with Magnaflux or electronic inspection to check for cracks in the tool joints. This will keep you out of trouble at all times.

16
Fishing Tools

A fishing job is one of the biggest problems a consultant will encounter when drilling a well. It is costly to the operator, hard on the hands, and especially hard on the consultant.

Once the cause of the problem is defined, the fishing tool company is called. One of the best around is Ponder Industries Inc. of Alice, Texas. I have used their hands several times and got great service. They will need to know:

- If the wellbore is cased or open hole
- Whether the fish is free or stuck
- The depth of the well
- The length of string in the hole
- The type of fish
- The OD and ID of the fish
- Whether it is drill pipe or a drill collar that has twisted off

Since you cannot know what condition the fish is in until you have it out of the hole or until you can examine the opposite break, assume the worst and have the fisherman bring out the necessary tools. There is no charge for the tools (except for freight); unless a special tool is actually made up, so it is best to have any tools that might be needed sent.

After the string is tripped, the damage can be assessed. If it is not too bad, the job should be simple. If the damage is severe, special tools may be needed. Successful fishing is a combination of the right tools, the right decision, and a fisherman who knows how to attach the fish and bring it to the surface. A general rule of thumb: When the cost of fishing approaches 50% of the cost to cement and side tract, the operator will usually go for it.

A simple fishing job typically requires a string of the following tools (see Figure 16-1):

- *Overshot* (Figure 16-2)—Used to go over the fish and apply a spiral or basket grapple. The milled teeth on the grapple bite into the fish and hold it tightly—and the fish is caught!
- *Bumper jar*—A slip sub that allows you to pull up or hit down on the fish. It allows you to sometimes jar the fish out without using hydraulic jars.
- *Hydraulic jar*—Jars the fish in an upward motion. To cock, the jar weight is set down on the fish. When it is pulled up, the jar goes off with the aid of the accelerator jar. The number of collars used depends on how much weight is needed to cock the jar.
- *Accelerator jar*—Used to give the hydraulic jar more hitting power and also to keep the jarring effect from damaging the drill string. Some consultants do not use the accelerator, but I think anything designed to help should be used.

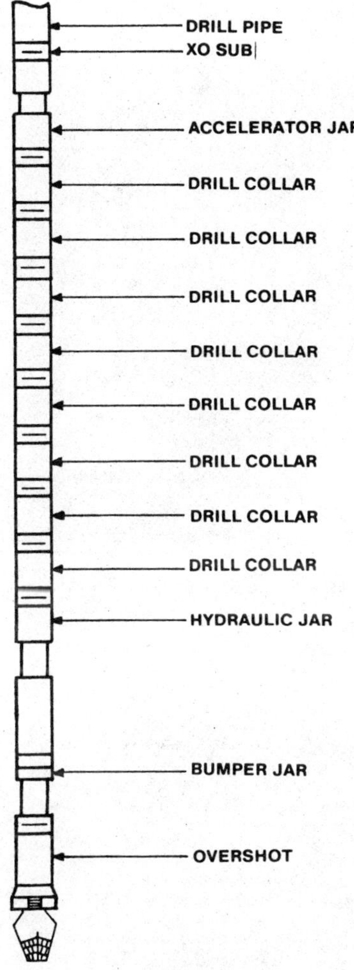

Figure 16-1. A simple fishing string.

Mills, cutters, and reamers (Figures 16-3 to 16-6), along with several hundred other tools, can be used for a fishing job when the situation arises. The mills are used to mill drill pipe and casing and to grind up trash in the hole. A junk basket should be run when mills or any tools are used that wash upward over the fish.

Figure 16-2. Overshots and tools. (Courtesy Ponder Industries Inc., Alice, TX)

Figure 16-3. Drilling jars. (Courtesy of Dailey Petroleum Services Inc., Houston, TX)

Figure 16-4. Milling equipment. (Courtesy of Ponder Industries Inc., Alice, TX)

Figure 16-5. Different combinations of mills and cutters. (Courtesy of Ponder Industries Inc., Alice, TX)

Figure 16-6. Reamers and special tools. (Courtesy of Ponder Industries Inc., Alice, TX)

On a simple fishing job, the string is lowered to about two or three feet from the fish, and then fluid is circulated to clean around the overshot. The string is then rotated to the right, and the overshot is lowered over the fish.

When a weight reduction is noted on the string, rotation is stopped and some weight is set down on the fish. When the fish is tagged, the grapple sets. If when the string is pulled up the weight on the string is equal to the original weight, the fish is caught. The string is not rotated on the way out since this would release the fish. Therefore, it is necessary to chain out the hole.

If the fish is not caught immediately, it could take hours of maneuvering to catch it. The fisherman should know how to approach the problem and what tools to order out if the overshot fails. If the fish is in a large washout area, a knuckle joint and wall hooks can be used.

Differential Stuck

Being ''differential stuck'' is one of the main causes of fishing jobs. It is caused by the hydrostatic pressure of the drilling fluid pushing the drill pipe against the wall while wall cake forms and creates friction great enough to cause the string to stick. Spiral collars are such a great asset in these instances because the spirals keep the wall cake from forming a friction seal against the collars.

The first thing most consultants try to do is get a free point and back off the pipe and go fishing. But before attempting that expensive project, you should order out a free point wireline unit to locate where the pipe is stuck. Then calculate how many barrels of Black Magic™ will be needed to pump around the stuck annulus portion of the pipe to eat up the filter cake [dissolve the mud on the sidewall (see Figure 16-7)]. (A chart of capacity and displacement of drill collars is included in Appendix E.)

Example

Your hole is 8,750 ft deep. After running a free point, you determine you are stuck at 8,627 ft. Other pertinent information is:

- Hole size—8¾ in.
- 7-in. drill collars—BHA 542 ft
- 4½-in. XO drill pipe—16.60 lb/ft

First find the capacity of the drill pipe and drill collars. The cement book will provide you with the following information:

  drill pipe capacity (bbl/ft) = 0.01422
  drill collar capacity (bbl/ft) = 0.0088

Using this simple calculation:

displacement volume = (DP bbl/ft)(DP length)
$$+ \text{(DC bbl/ft)(DC length)}$$
$$= 0.01422 \times 8.208$$
$$+ 0.0088 \times 542$$

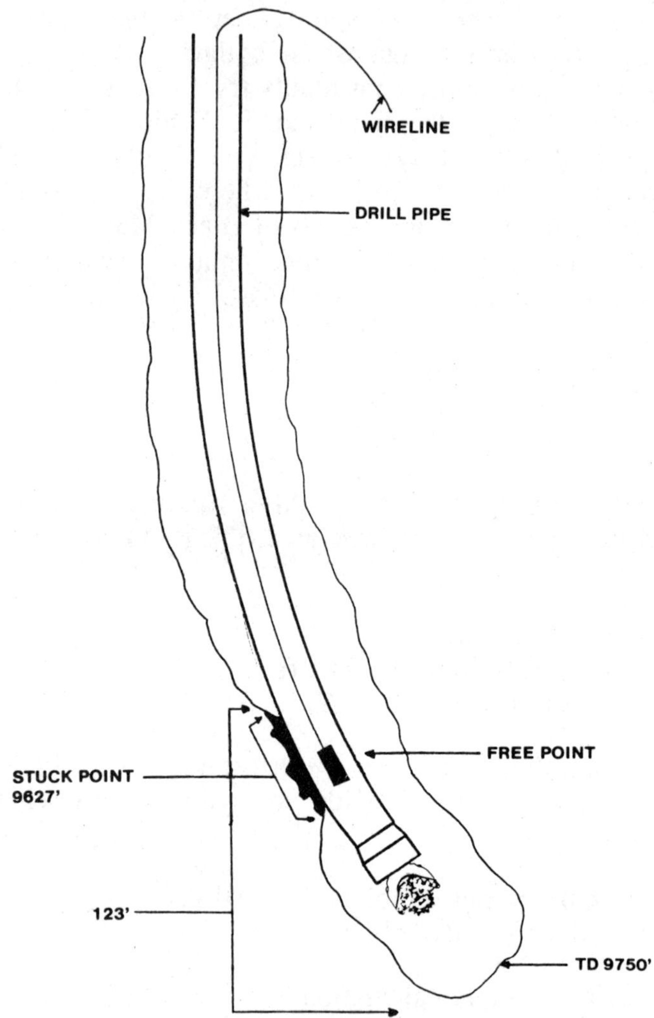

Figure 16-7. A pipe "differential stuck."

$$= 116.71 \text{ bbl DP}$$
$$+ \ 4.76 \text{ bbl DC}$$
$$= 121.47 \text{ bbl } (121.5 \text{ bbl}$$
$$\text{rounded off)}$$

Next find out how many barrels from the bit to the stuck point. Using volume and height between hole and drill collars, we use the annular volume between 7-in. casing and 8¾-in. open hole for the drill collars which is

$$0.0268 \text{ bbl per foot} \times 123 \text{ ft} = 3.29 \text{ bbl}$$

or rounded off to 4 bbl. (In the oil field you always round off to the next largest whole number when figuring outside annular capacities, because of the washout factor. Petroleum engineering calculations are not as accurate in the annular portion of the open hole.)

Since the number of barrels to stuck point is small, figure 100% excess, which is 8 barrels. Since you need to move one barrel Black Magic™ every hour, figure 40 more barrels to cover the displacement for 40 hours, so the total would be 48 bbl. Round this off to 50 bbl, since you have to order Black Magic in 5-barrel increments.

Order out 50 bbl of Black Magic™ at the same weight as the present drilling mud and a pump truck to pump the fluid.

While this is on the way, figure how many barrels are needed to pump down the drill string in order to displace fluid to the stuck point, leaving enough fluid in the drill pipe to pump one barrel per hour. This will keep the area near the stuck pipe saturated with Black Magic.™ The capacity of the drill stem is 121.5 bbl and the volume to the stuck point is 8 bbl rounded off plus 2 bbl excess when ordering, figuring 100% excess on the outside. So the total is 131.5 bbl.

Since we want 40 bbl of Black Magic™ in the string, we will only pump 131.5 bbl of fluid. We lead with 50 bbl of Black Magic™ and tail with 81.5 bbl of mud. That will pump 10 bbl around the annulus of the pipe.

Black Magic™ takes 30 to 36 hours to work, so you have to wait. Don't make the mistake of trying to free the pipe too soon. This approach takes time, so do not even touch the pipe for at least 12 hours. After that attempt to turn the pipe every hour. On the Gulf Coast, it takes an average of 32 hours for the filter cake to dissolve. Even though the operator will want quick miracles, you should make him be patient. This procedure will work just about every time if it is given a chance.

Every hour pump one barrel of fluid down the drill pipe and check the surface to see if one barrel comes back up over the shale shaker. This will ensure a moving fluid and a solid hole (no washouts). Put six to eight turns on the drill pipe at the floor, and latch the tongs to the pipe to hold the torque. Then work the pipe up and down to work the torque down the hole. This will sometimes roll the pipe out of the keyseat or stuck area.

Remember that when you have a stuck pipe, you are not in terrible trouble if you have circulation. The Black Magic™ will work; the biggest problem is being patient while it does. If the jar will go off, keep using it while working the pipe—you may get lucky.

If the dissolvent does not work, run a free point and back off the pipe two joints above the stuck point, then go after it with an overshot and jar it out. Most of the time this will work but sometimes you cannot get to the fish. If not, you have to order out wash pipe to wash over the fish. This is a very expensive project and one that just adds to the problems (see Figure 16-8).

Wash pipe can be run several different ways. You can burn over the fish and clean it out for 80 to 500 ft, then pull out of the hole and go after it with a screw in sub or overshot. If that does not pull the pipe free, you can back off above the new stuck point and continue the washing procedure until you get all the fish. Also, with wash pipe, you can run a backoff connection tool which allows you to get the fish with the wash pipe in the hole. Then trip out of the hole (this takes five or six hours with 400 to 500 ft of wash pipe).

Fishing is a complicated procedure with many different types of tools for freeing and retrieving fish. Many fishing tools have to be designed for the job. If the job becomes complicated, the fisherman can explain what he is going to do, or if the expense becomes too great, it may be necessary to order the BHA left in the hole. In this case back off as deep as possible, set a cement plug, kick off, and drill a new hole to TD. This is expensive, but sometimes it is a necessity to continue drilling the well (see Figure 16-9).

(H) In horizontal drilling most fishing jobs arise when cones have been left in the hole and motors whiplash off the horizontal string. There is very little pipe separation, because the pipe is not rotated very quickly and often you slide the pipe. Many operators have reentered the older chalk wells to cut a section or window and kick off and begin horizontal drilling. Ponder Industries Inc. has a good section milling tool for 15 to 20 ft and a new section mill capable of going 60 to 90 ft (see Figure 16-10). Many of the old chalk wells drilled during the boom, back in the 1970s, are candidates for horizontal drilling. The cost is much less because the vertical hole has already been drilled and encased.

The worst problem is having to fish a motor out of a horizontal well if the well is flowing. This could be a problem if heavy brine water had to be used. Heavy brine water, in my opinion, kills off half the well, but in a situation where you need to fish a well, it may be necessary.

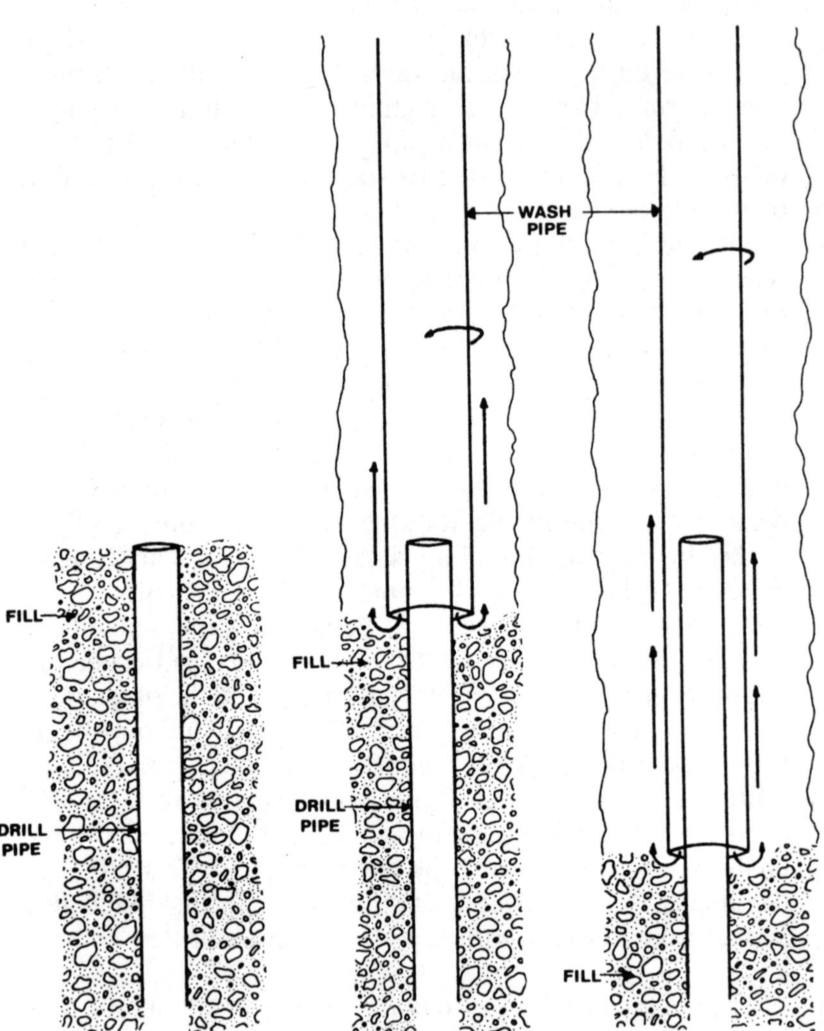

Figure 16-8. A wash pipe operation.

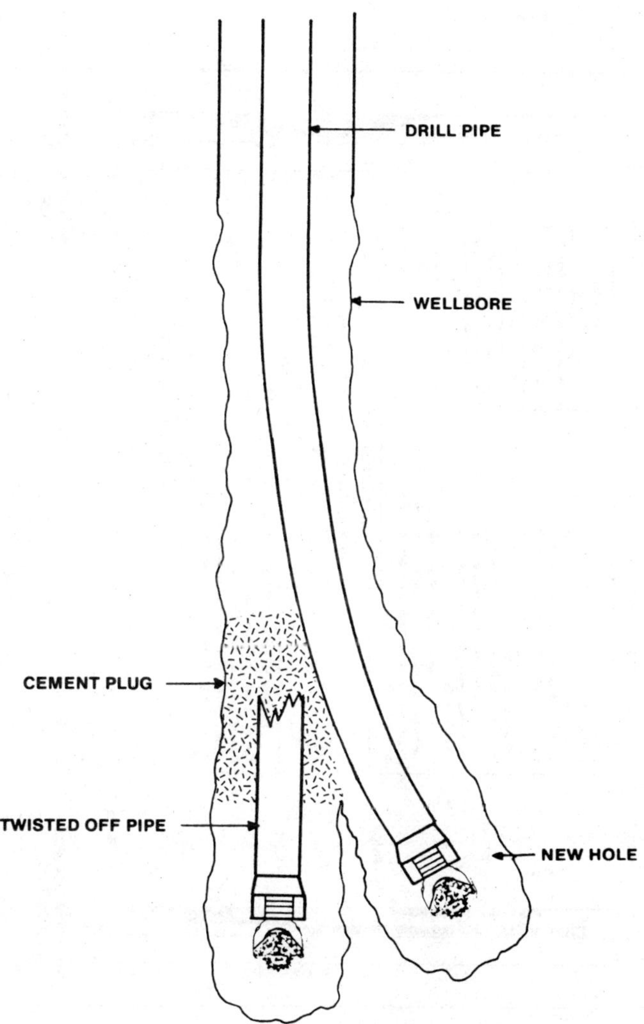

Figure 16-9. Setting a cement plug and kicking off to make a new hole.

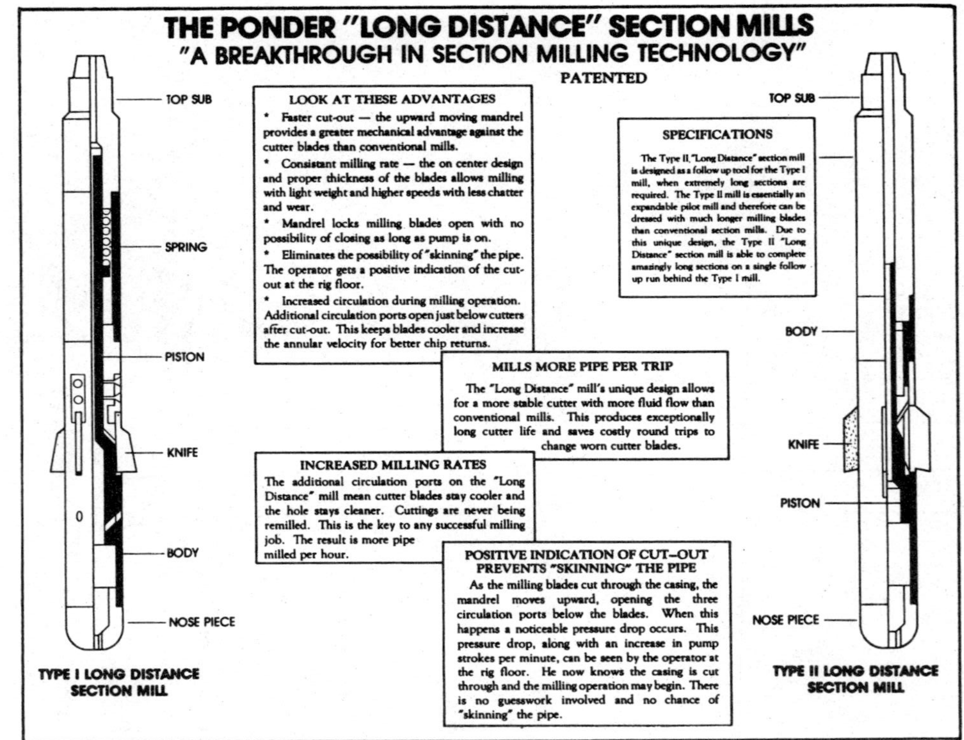

Figure 16-10. Section milling tool. (Courtesy of Ponder Fishing Tools, Inc., a division of Ponder Industries Inc., Alice, TX)

17
Drill Stem Tests

This test is used to make a temporary completion on a well. The purpose of the test is to determine the bottom hole pressure (BHP) and to give the engineers more detail on potential oil- or gas-producing zones. Because of the many problems associated with running drill stem tests (DSTs), they can be a real headache for consultants. A few of the problems include: packers getting stuck, rubber packer skins being left in the hole, test failure, leaking or damaged packers, the possibility of blowouts, and hydrogen sulfide hazards.

The order to run a DST will usually come from the geologist on location when a show of gas or oil is encountered. The show may be from cutting samples, from the gas unit log, or from a core barrel sample that has been run in an oil-producing zone.

The geologist or testers are responsible for evaluating the results of the test, but it is the drilling consultant's job to *get* the results—a much more difficult task. The consultant must condition the hole and the mud, make some short trips to ready the wellbore for the DST tools, and remove the tools from the hole after the test. After the geologist has left, the consultant may still be working to pull the tools from the hole, which is sometimes complicated.

When the order is given to run a DST, call up a tester and give them the following information:

1. Hole size
2. Depth
3. Rat hole depth (the depth to which tests will be run below the zone)
4. Basic hole condition
5. The kind of tools to bring (this specification will come from the geologist or the operator)

There are many combinations of DST tools available, including:

- *The straddle packer*—This setup requires two inflatable packers that are filled by rotating the surface pipe. The straddle packer test is used to isolate the bottom zone from the higher test zone. Usually it is run if the geologist misses a pay zone and the logs have been run. A DST is run to evaluate the well before it is plugged or pipe is run. Normally, two or three recorders are run with this method (see Figure 17-1).
- *The standard packer*—This is used more often than any other method. One packer and an anchor are used. The packer is run above the test zone with an anchor pipe to take the weight of the string and expand the packer to seal above the zone at the bottom. Most people run two packers just in case one packer does not hold, because using two packers is much cheaper than rerunning the test. Normally two or three recorders are run with the test (see Figure 17-2).

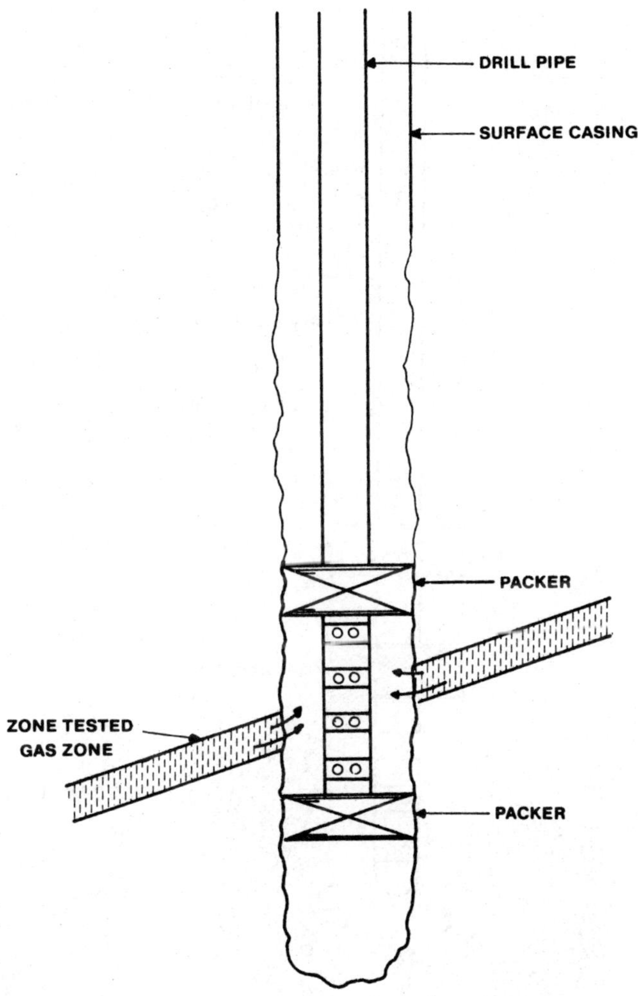

Figure 17-1. A straddle packer.

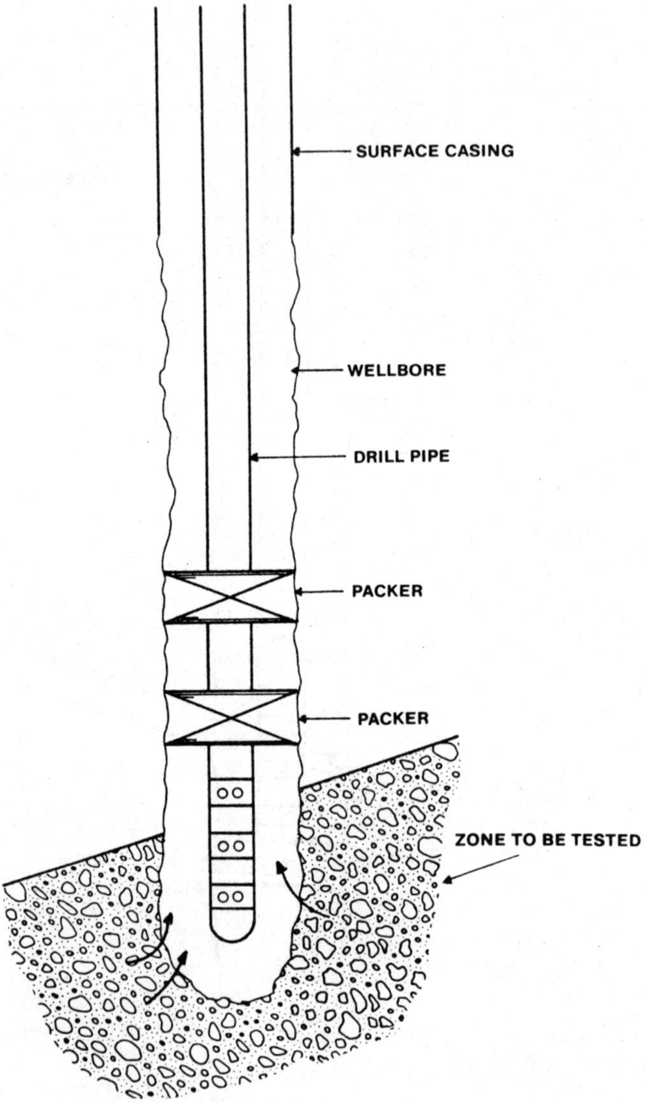

Figure 17-2. The standard drill stem test.

- *Cone packer*—This is a packer that can be set down above a core hole to test the hole for pressure. All you have to do to set the packer is put weight on it, and it will seal the area over the hole. You can also run a packer above the cone packer to assure a seal. (See Figure 17-3.)

The DST is actually a simple concept. The pressure and time of tests are recorded, and samples of the formation fluid are collected for the geologist. The recorders are part of the DST tool. Most people run two recorders in case one set is plugged. The testing company's representative will advise which recorder to use. In the test, the following information is recorded:

1. Initial hydrostatic pressure exerted by the mud column (IHP)
2. Initial closed-in pressure (ICIP)
3. Initial flowing pressure (the lowest pressure recorded just after the tool is opened) (IFP)
4. Final flowing pressure (the pressure just before the tool is closed) (FFP)
5. Final closed-in pressure (FCIP)
6. Final hydrostatic pressure (FHP)

A sample catcher that collects some of the formation fluid for evaluation at the surface is also part of the setup. The fluid can be sent to a lab to determine if hydrocarbons are present. If a shale-healing asphalt-base chemical is being used in the mud, send a sample of it to the lab also.

The recorders show what pressures are present downhole as well as pressure increases during flow and shut-in periods. A consultant is not required to read the recorder results, but with a basic knowledge you can know what is going on. (See Figure 17-4.)

The flow period and shut-in period will be determined

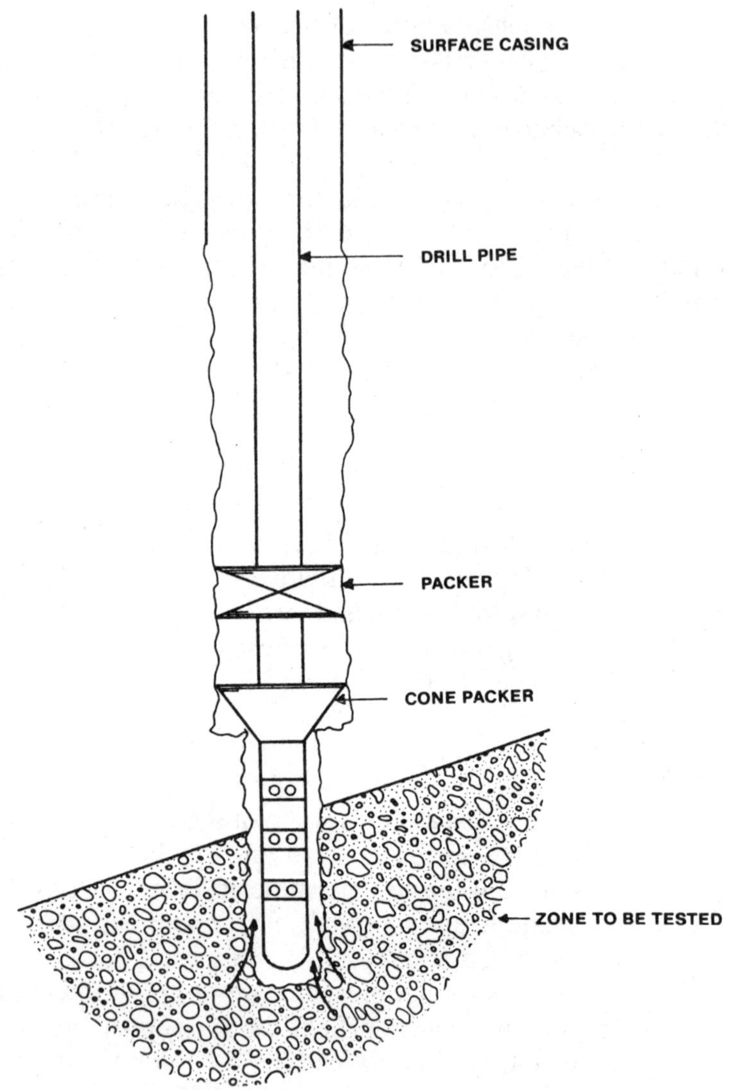

Figure 17-3. The cone packer test.

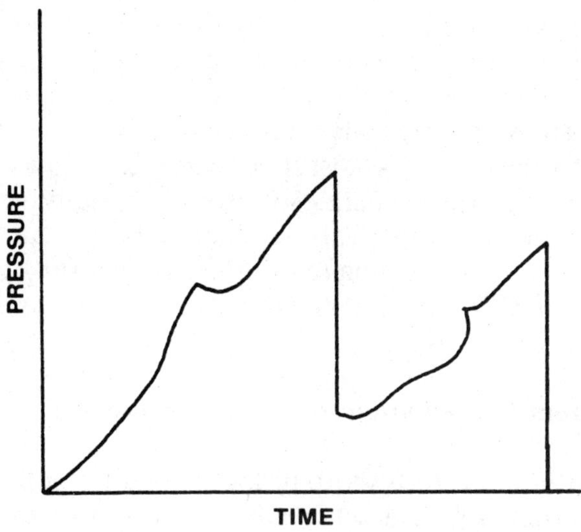

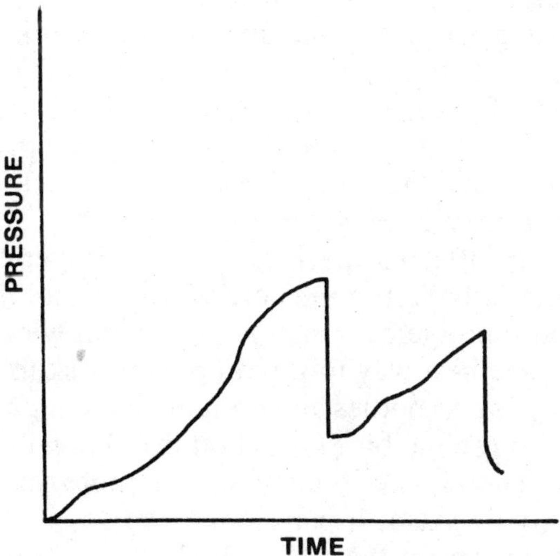

Figure 17-4. DST pressure chart.

by the geologist on location or by the operator. Remember in bad hole areas it becomes more critical to remove the tool the longer you stay on bottom. In hard rock areas longer periods are no problem. Because of bad hole conditions on the Gulf Coast, use less test time. Some state laws prohibit pulling the tool after dark, so tests are usually scheduled early in the morning. This regulation is due to fire and blowout hazards encountered during a DST run. *Caution: Never pull DST packers at night because of possible fire hazards.*

Problems Encountered When Pulling DST Tools

When you are ready to pull the test tools, pull them up with caution. Fill can settle around the top of the packer and form a seal. If this happens, pull up about 50,000 lb over string weight and hold it for a few seconds, then let it down slowly. Repeat the operation a few times. This will usually break the packer seal. If that does not work, gradually increasing the pull sometimes works. Work with the packer for a while before you decide to back off the safety joint and settle for a fishing job. Sometimes equalizing will break the suction of the packer. Most testing companies will not run a jar above the tool. If you have to back off, you will need a fisherman to come out and get the fish.

When using inflatable packers, a high temperature will cause the skin to be left in the hole when the tool is pulled. This will cause problems if another test is run with the skin in the hole. The best way to grind up packer skins is break up a dozen soda pop bottles and drop the broken glass down the well. Then use the bit to grind on the skin with the pop bottle glass. The glass will chew up the rubber better than the bit alone will. Also, if permission can be obtained, run a Visbestos sweep on the hole. (Visbestos is an asbestos additive that cleans the hole by increasing the viscosity.)

This will bring large pieces of rubber to the surface. Although Visbestos has been outlawed in many areas because of its harmful effects on people, using the right breathing apparatus will make it safe. To sweep the hole with Visbestos, simply pump it down the hole mixing it for 10 to 15 minutes. This will efficiently sweep the hole of debris.

Water Cushions

It is important to run a water cushion (by running water inside the pipe during a DST) to keep the drill pipe from collapsing down the hole. It is very important to keep the water cushion underbalanced so that the test will have a chance to record the data for the geologists. The tester will be able to give you some idea of the best length of the cushion.

(H) In horizontal wells there really is no need for a drill stem test. However, once horizontal wells began to be drilled in higher-pressure sands and with mud, it became possible to take a drill stem test. In some instances it would be cheaper to find out if the well were commercial before wasting the running of liners, etc.

18
Coring the Well

This is a simple procedure that allows the operator to evaluate the geology of the well more accurately. It requires obtaining a 3.5-in. diameter core from the well and analyzing it in a laboratory.

The conventional coring method is used more than wireline coring in the oil field, because the former method yields a larger sample size. However, to retrieve a core sample in the conventional way, you must make a trip to the surface, which is time consuming. The wireline method will become more popular as consultants learn that it is easier to operate.

The conventional method uses two kinds of coring heads:

1. A roller-and-cutter head (standard)
2. A diamond head

The roller-and-cutter core heads are only good for 25 to 30 ft, whereas the diamond core head is good for up to 55 ft. The diamond core head will retrieve more length if necessary, and that, in turn, will save you money. If just a small core is needed, the cheaper roller-and-cutter heads are better.

Wireline coring is beneficial only in deeper wells. The wireline is run down the drill pipe and catches the core barrel with an overshot that allows you to bring it to the surface without tripping.

Normally a service hand is on the rig to make up the barrel for the run. He will be able to explain the operating procedure and weight needed for the operation. After the core barrel is made up, it is tripped in the hole slowly to ensure it does not hit the sidewall or dog legs, as this would fill the barrel before it reached bottom. *Take your time.* Start coring with only a little weight until you get some penetration and then gradually add more weight. Also keep the rpm down until you establish a rate of penetration.

Keep an eye on the pump pressure. When you are on the bottom, the pump pressure will be greater. When you pull off the bottom, it will decrease. The increase in pump pressure will keep the core head clean. If the pump pressure goes up and the core head is still not on bottom, trip the core head out from the hole and check to see if it is already full. (See Figure 18-1.) The core barrel could also be full or plugged if the rate of penetration decreases. Accurate depth measurements are important in coring operations. The driller should watch for all the signs of a filled barrel and changes in pump pressures to avoid tool damage.

When the core barrel is pulled, it should be done slowly to avoid sucking the core plug back out of the barrel. This is very important for the success of the test. Once the core sample is on the surface, the geologist will take over.

The sidewall core gun is the easiest method to use. The gun is run into the hole with a wireline unit. Quite a few samples at different depths can be obtained with this tool. Most guns come with 40 shots. (See Figure 18-2.)

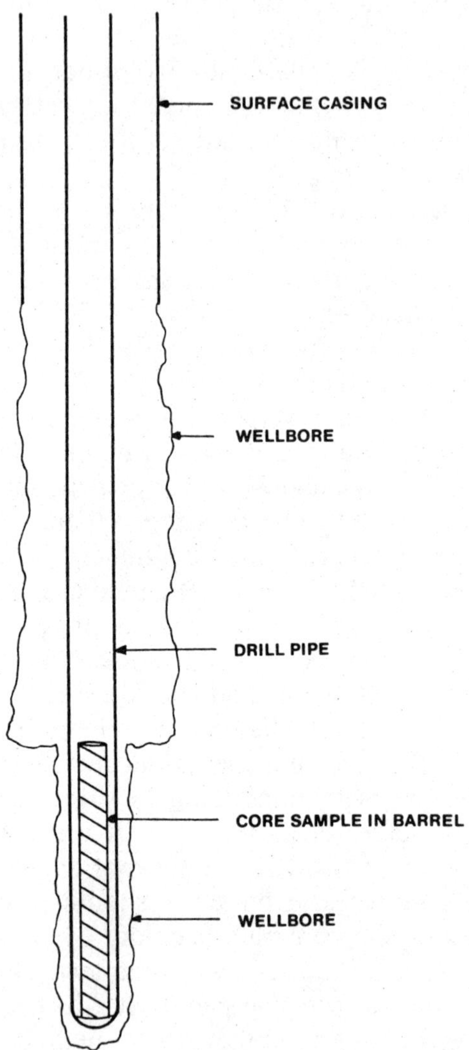

Figure 18-1. Taking a core barrel sample.

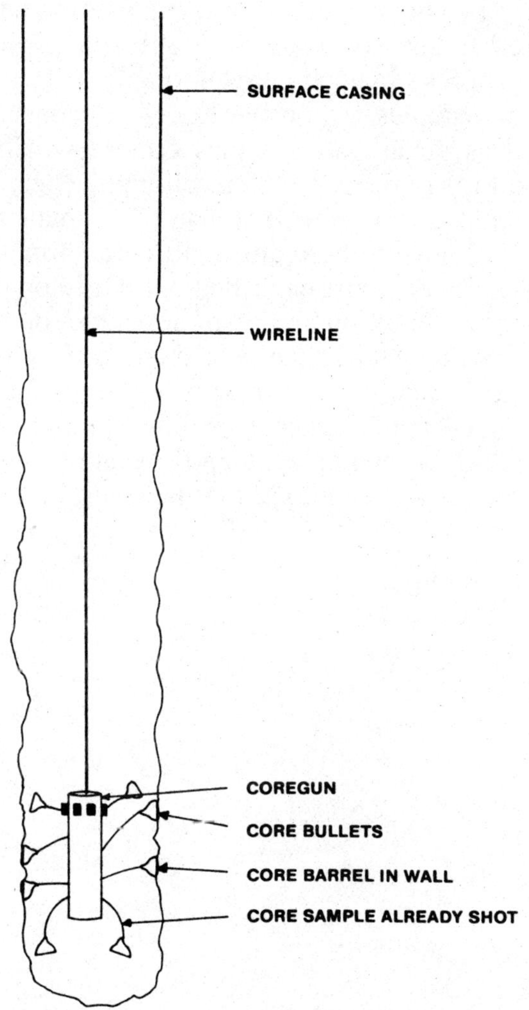

Figure 18-2. Using the side wall core gun to gather samples.

The biggest disadvantage to the sidewall core gun is that core samples are small. However, if the samples are wrapped in plastic and stored in airtight containers, and in some cases frozen, they are as usable for evaluating the geology as cores obtained conventionally.

Once the samples are on the floor, the consultant's job is finished as far as core samples go. He will next either run a DST or drill deeper. When drilling deeper, it must be remembered that the core barrel hole is smaller than the core and will have to be reamed. In hard formation areas, the core barrel hole must be drilled out. Once on the bottom of the core hole, pull up and ream down two or three times to widen the diameter of the hole then go back to drilling.

(H) In horizontal wells, coring is not needed at this time; however, once the horizontal drilling begins in hard formations and tight sand areas, then some application may be used. This is new technology that needs to be developed.

19
Logging the Well and Accompanying Problems

When you "TD a well" (drill to total depth), the next step is to "log the hole" with electrical, nuclear, and sidewall core samples. The purpose of logging is to determine the basic characteristics of various potential pay zones in the borehole.

Oil and gas are not found in pools or caverns but are dispersed throughout the reservoir rock. Not all rocks will hold fluids. In order to hold oil or gas, the rock must have spaces between the rock granules. The ability of rock to contain fluids is measured by its open pore space and is called "porosity." It is measured in percentages. If the porosity is 20%, it means that 20% of the rock volume is pore space available for fluids and 80% is solid rock.

It is not enough for a rock to contain oil or gas. In order for oil or gas to be recoverable, it must be able to flow into the wellbore. The fluids can flow only if pore spaces are connected. The ability to flow is called "permeability" and is measured in millidarcies, often abbreviated md.

The ability of oil or gas to flow also depends on the pressure in the zone, which provides the driving force to push the fluids into the well. It is important in analyzing the potential zone to know this pressure. This is measured in pounds per square inch, abbreviated psi.

It is also important to know the type of fluids in a zone— whether they are hydrocarbons or salt water. Salt water is a good conductor of electricity, and hydrocarbons have a high resistance to electricity. Therefore, tests are run to determine the "conductivity" and "resistivity" of a formation. High conductivity indicates salt water and high resistivity indicates oil and/or gas.

Since the consultant is not hired to analyze the logs and samples, you should not make any suggestions unless you are trained in log analysis. Your only concern is to get the logging device to the bottom and back up again. You are paid to drill a hole, not to analyze samples. This chapter will touch on the kinds of logs there are, how they work, the problems of running a log, and in particular, the problems of getting stuck in the hole. There are many combinations of logs to run and different types of data to analyze.

Induction Tools

The first logs run are generally run with induction tools. These tools measure the resistivity of the formations and thereby determine the location of hydrocarbons in the well. They will operate in temperatures of up to 350°F and pressures of up to 15,000 psi.

Porosity-Lithology Logs

These logs are the density, sonic, gamma ray, and neutron logs. Hydrocarbons affect these logs and are recorded on the drilling log. These tools are rated at 350°F and are used in most wells.

These logs measure the amount of pore space in the hydrocarbon-bearing formations. From this the geologist can estimate the number of barrels in the reservoir thereby deciding if the well has commercial potential. These logs are run by the sonic, density, and neutron logs. The density and neutron tools have radioactive sources on them, and much care should be taken not to get them stuck, since the federal government may require their removal.

Dipmeter and Directional Logs

The dipmeter is run to give a correlation to other wells and to determine the formation dip. The survey part of this tool will give hole deviation and true vertical depth (two). A caliper survey is also part of the system and is used to determine cement values (amount of washout in borehole).

When the logging truck arrives on location, be sure the log engineer has the correct tools. Before any operation starts, sign for tool protection insurance. This can be important if the tool gets stuck and cannot be fished out. The tool protection insurance costs about $40 compared with the $30,000 to $40,000 cost of some tools. With insurance you are usually required two tries to fish a tool after which you have the choice of continuing to fish or of setting a plug, kicking off, and drilling a new hole.

In some cases the U.S. Geological Survey requires that nuclear logs be fished out, which could become quite expensive to the operator.

The consultant should be on the log truck until the tool reaches the bottom; then his job is done. Going to the bottom with the log can be difficult sometimes because of bridges and trash in the hole. *It is important to clean and condition the hole and the mud before the logs are run.*

If the tools hit a bridge you should try to work them through it by playing them up and down. If this does not work the tools through the bridge, you will have to pull them and go back in the hole with the bit and clean out the hole. This can be quite expensive and time-consuming.

If the tool sticks coming out of the hole, have the engineer inform you before he pulls too hard on the tool. The tool has a special pull-off socket that will allow the wire to come off the tool, leaving the tool in the hole.

Before you pull off the tool, put some pull on the tool and let it set. Sometimes the pull will free the tool. Never pull off the tool until the office is notified. Put some tension on the tool and call in. Sometimes by the time you call, the tool will have come loose and the cable and tool will be on their way to the surface.

If the decision is made to pull off the tool, put maximum pull weight on the tool and call a fisherman. Find out how long it will take for him reach your location and then go work with the tool. Pull off the tool just before the fisherman arrives and get ready for him.

Since the tool has a special surface for fishing, fishing usually is no problem. Sometimes there is a problem getting over the top of the tool if it is against the wall. When this happens, special hook wall joints will be needed to get the fish. Another system of fishing is to use the wire to guide the fishing tools over the fish. This method is expensive because the wireline must be cut.

Most tools get stuck because the tool fails to close properly and one of the pads becomes stuck against the wall. The tool man will sometimes try to blame the stuck tool on the hole conditions, but before you pay for the run, examine the tool. If the pads are not all in the closed position, then tool failure caused the tool to stick and you should not pay for the misrun.

Although most logging companies keep their tools in good condition, many still fail. If a tool has not been reconditioned, it is especially subject to failure. Logging failures are common so expect them, and do not let the logging engineers put the blame on hole conditions. Always look their tools over carefully.

In the horizontal well logs are run to correlate between wells in the area to find the right spot to begin horizontal drilling. Another log that is used by some geologists is the *cibal log*. It shows fractures in the wellbore, which lets the geologist determine the best entry spot and gives him an idea of how well the well is fractured. The cibal is very expensive, and some operators do not use it.

20
The Intermediate String, Liners, and Testing

The intermediate string is a protective casing run in a well when hole pressures and hole conditions merit it. This casing is run in the same manner as surface casing, and the operation is simple.

After it is determined that an intermediate string is needed (or it is called for in the prognosis), strap and prepare the casing for running. Remember to strap and tally the casing personally to reduce the possibility of mistakes. Calculate the total length of casing on location to determine which joints to take out, just as you did with the surface casing.

Example

The intermediate string will be set at 8,200 ft, and there are 8,292 ft of casing on location. Joint 21 is 42.92 ft, and joint 39 is 44.19 ft. By taking these two joints out and subtracting their total length from the string, you are left with 8,204.89 ft. Add to this amount 1 ft for the guide shoe and 3.72 ft for the float collar for a total length of 8,209.61 ft.

Have the driller drill to 8,207 ft. This will leave 3.61 ft above the rotary table, which is a perfect height for placing the cement manifold. This will also allow the casing to be set 1 ft off bottom. Have the casers and cement crews arrive on location by the time you come out of the hole. Rig up the casers and run the casing. Remember to fill the casing every five joints with drilling fluid.

Though the examples in this book call for only one type of cement, both a lead and tail cement are always required. The lead cement is composed of a light cement and the tail a neat cement (neat meaning heavy). Neat cement is better to put around the shoe, so that the shoe will test more easily when the EMW is tested.

Check with the cement engineer to make sure the cement is the right formula. Since the hole is 8,207 ft deep and you need to cement up to 4,500 ft, you need to determine the number of sacks needed for 3,707 ft of cement. First check the cement book for:

1. Volume between casing and open hole—The casing is 7 in. 23 lb/ft, and the hole 9⅞ in., so we set the formula:

    $$0.2647 \text{ ft}^3/\text{ft} \times 3,707 \text{ ft}$$
    $$= 981.24 \text{ ft}^3 \text{ annular capacity}$$

2. Next find the capacity of the casing to float collar. Since the float collar is at 46 ft, the casing capacity to float collar is

    $$(8,207 - 46) \times 0.0394 \text{ bbl/ft}$$
    $$= 321.54 \text{ bbl } (321.5 \text{ bbl rounded off})$$

    needed to displace cement with mud down to the float collar.

3. The cement capacity of 46 ft of 7-in. 23 lb casing below the float collar is:

$$0.2210 \text{ ft}^3/\text{ft} \times 46 \text{ in.} = 10.16 \text{ ft}^3$$

The total annular and casing cement capacity is:

$$981.24 + 10.16 = 991.4 \text{ ft}^3$$

4. To convert cubic feet of cement to barrels of cement adding 20% excess, we perform the following operation:

$$991.4 \times 0.20 = 198.28 \text{ ft}^3$$

$$198.28 + 991.4 = 1{,}189.68 \text{ ft}^3$$

Converting to barrels use the constant:

$$0.1781 \times 1{,}189.68 = 211.80 \text{ bbl cement}$$

5. To find sacks use the yield of 1.20 cubic feet per sack

$$\frac{1{,}189.68 \text{ ft}^3}{1.20 \text{ ft}^3/\text{sack}} = 991.4 \text{ sacks}$$

Cement excess is determined by the cement company, since they know what is used in the area of the well. Excess can vary from 20% to 100% depending on geology in the area. The Gulf Coast area is famous for much excess, so check with the cement company before cementing the hole.

When the casers finish running the casing, tag bottom, pull off 1 ft, and set the slips. Rig down the casers and rig up the cement manifold. Break circulation and get bottoms up to get possible gas or fluid invasion out of the wellbore.

Now cement the well. Pump the 211.8 bbl (20% excess) of cement and insert the plug. Follow the plug with 321.5 bbl of mud to displace the cement to the float collar. Watch for an increase in pump pressure when the plug is bumped. Record that pressure and the time. Check for backflow. If there is backflow, repump the regained fluid down the hole and close the cement manifold at the surface. Some operators require that the pipe be reciprocated (moved) before the pipe is latched down to set. That is up to the office engineer. I sometimes move the pipe 5 to 6 ft to position the cement all around the pipe. This is one area where no one is an expert. We do not have downhole glasses. Now wait on the cement for 12 to 18 hours (see Figures 20-1 and 20-2).

After the cement job is finished on the intermediate string and the cement is dry, the next step is to cut the casing and tie into the system.

Trip back down the hole, strapping pipe as you go, and tag the cement. Record that depth to make sure the calculations are correct. Close the annular preventer and pressure-up the system to 800 psi to check for leaks in the casing. If no leaks are found, drill out the float collar and retest for leaks. Drill out the guide shoes and 10 ft of formation, and test the new shoe to whatever the prognosis calls for. If everything is all right, go back to drilling.

Hanging a Liner

This is a method of saving money by eliminating the need to run casing from the surface to TD. The running of a liner is sometimes used to TD a well and protect the wellbore while drilling deeper, or when abnormal pressure is encountered and casing is needed to keep the intermediate shoe from blowing out. In the following example the hole is too hard to control and you are short 700 ft from TD.

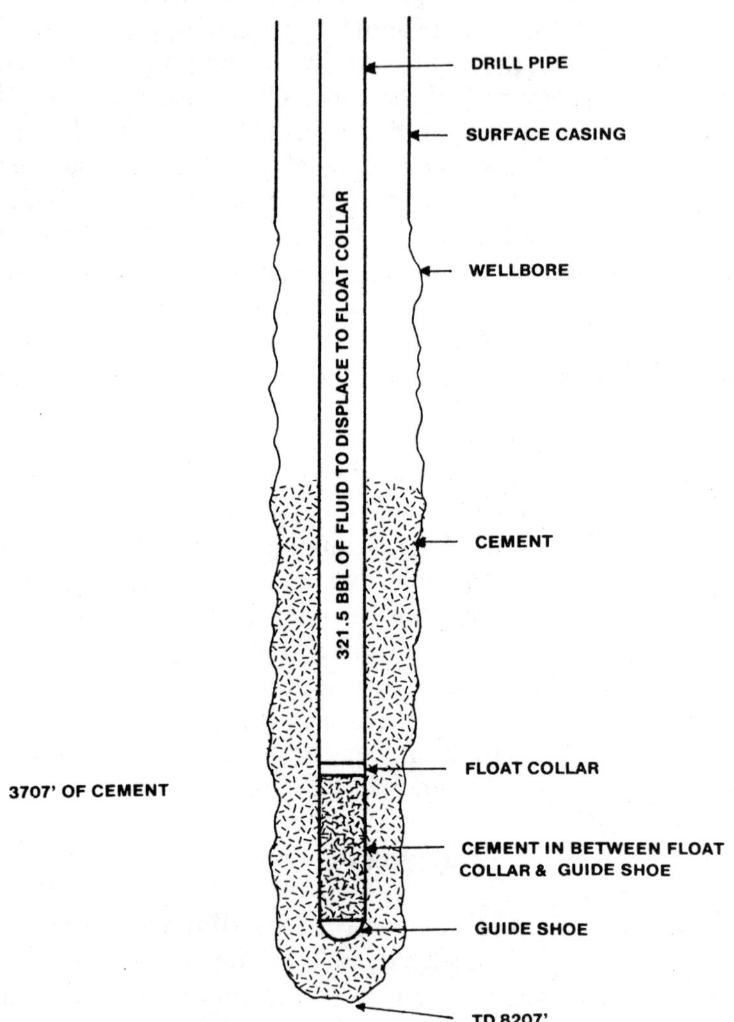

Figure 20-1. Cementing the intermediate string.

Figure 20-2. Cement a 7-in. string in the vertical hole of a horizontal well. (Courtesy of Davenport Horizontal Drilling Consultants, San Antonio, TX)

Instead of running a long string and then a liner hanger to TD, run a hanger to 8,500 ft, drill out, and finish the well. Since at this point you really do not know if you have made a well, it is easier and cheaper to hang a liner and, after running the logs, to decide if another liner hanger needs to be set to TD or if a small production string should be set. Hanging a liner is usually a last resort when trouble occurs.

Example

You are drilling at 8,500 ft. You are having a hard time controlling the shale and you need to protect the wellbore from sloughing, so the engineer orders a liner run. Since you have a 7-in. 23 lb/ft casing to 8,207 ft, order out a 4½-in. 16 ft casing liner.

Order a liner hanger tool, a serviceman to run the job, the cement, a cement engineer, and a casing crew. The cement engineer will calculate how much cement will need to be run. Tell the casing crew the size and weight per foot of casing needed and the depth it will be run.

When the casing arrives on location, be sure to tally it personally. You need enough length to leave 50 ft of liner in the intermediate casing. So if the intermediate casing is to 8,207 ft, 8,257 ft of liner casing will be needed.

If, for example, 8,298 ft of liner were delivered out, and we take out joint No. 26, which is 42.61 ft, we have 8,255.39 ft. Add 1 ft for the guide shoe and 3.72 ft for the float collar, giving 8,259.11 ft, which is perfect for running the liner. Add the length of the liner hanger, which is 10.20 ft for a total of 8,269.31 ft.

When the liner man arrives on location, tell him what you need to do. He will take all the measurements and put the liner hanger with the casing in the exact location you need it. Rig up the casers and run the casing into the hole. Use slimhole centralizers when running a liner since the liner is normally a close tolerance to the casing to be hung in. The liner will be run the rest of the way with drill pipe.

Once the liner hanger tool is in place and cemented, it will set the liner and release the drill pipe. This is a very simple operation; however if you release the casers before the liner reaches bottom, the liner may not get to bottom, so you may have to come out of the hole and break out the

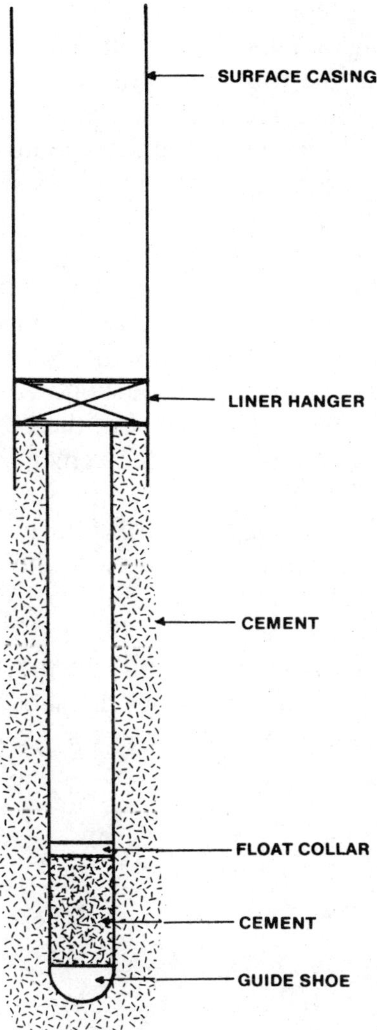

Figure 20-3. Hanging a liner.

casing so you can clean the hole out. This does happen on occasion, so be prepared for the worst.

The cementing of a liner is very important because a good seal is needed at the top to prevent leaks. On cement calculations, figure 20% excess and try to get your calculations on the money. The cement should come up the liner on the annulus side, yielding a good bond and seal between the liner and the surface casing.

Once the cement is pumped, displace the cement to the float collar and bump the plug. As the cement reaches the top of the liner hanger, an increase in pressure should be noted at the surface. This means the cement is squeezing in place and should make a good bond. After the cement job is finished, release the tool. Let the cement set for 18 to 24 hours. It is important that the cement has time to dry. (See Figure 20-3.)

After 24 hours has elapsed, drill out the casing and test the liner top. Trip into the hole right to the liner top. Close the annular preventer and pressure-up to 1,000 psi. If there is pressure loss, a squeeze job will need to be run on the liner top. If there is no loss, drill out the float collar and retest to check for leaks. Drill out the shoe and test for the EMW that the engineer calls for just as you would for any other string.

21
Finding the Horizontal Zone to Drill

When a horizontal well is being drilled, finding the right zone to drill is very important (see Figure 21-1). This is done in several ways. After logging the intermediate hole, some operators just correlate with logs from other wells in the area. This is cheaper, but not as accurate, and in some cases it causes problems. The best way to find the zone is to drill below the estimated zone depth and then run a set of logs to pick the tops of the zones.

The procedure is as follows: First, after nippling the stack, pick up the drill string and follow steps 36 to 42 of the procedures for horizontal drilling (in Chapter 4). After the well is logged, a zone is chosen and a special kickoff plug is set. This plus has some sand in it, so the curve-building tool will slide off the plug. Make sure that the directional drilling hand knows how to tally pipe correctly. That is his job, but I have seen experts make a 30-ft mistake. Before he drills, check his figures and make sure you both come up with the same number. If a mistake is made, another

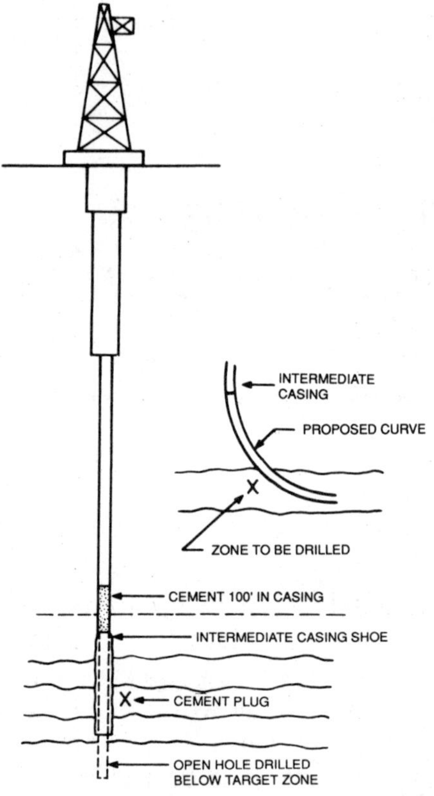

Figure 21-1. Finding the right zone to drill for a horizontal well.

plug will have to be set, and so it will be another 48 hours before anything can be done. Directional drillers make mistakes all the time. You have to choose them carefully. Make sure they have drilled several successful wells.

The hole to be logged can be drilled with water or field brine. Be prepared to fight a kick because you could drill into a pressurized zone while drilling vertical. This does not happen often, but a smart consultant will have the crews trained and ready to work the PWD equipment at this point.

Also you can drill the vertical hole with the cuttings going through the chokes and to the separation tanks. This also will teach the crews how to use the equipment before horizontal drilling is begun. A show of oil will probably appear. Before you withdraw from the hole, shut down the pumps and check for a small flow. If a small flow occurs, you can add to the system a little heavier brine water. This can be done by slowly opening the brine tanks to the mud tanks and mixing a heavier-weight brine water. Always be careful when you are in the chalk. After you are in the casing, also check for flow. Keep the water flowing to fill the hole while pipe is pulled. Let the overflow go through the chokes, but kick out the pumps every 1,000 ft and check for flow.

22
The Horizontal Directional Driller and MWD Tools

One thing essential to horizontal drilling is the development of horizontal drilling tools. Directional drillers have been used offshore for years, so the basic technology has been available. The first horizontal well really shook up most old-timers. The thought of horizontal drilling seemed ludicrous. Most people did not realize that the horizontal well had a large curve before it became flat (another word used to mean horizontal). The area of the curve is large enough that the drill pipe will bend around the curve and can also be rotated.

The directional driller normally works for a company that furnishes tools for the project. The day-rate cost is very high. Directional drillers are now in such demand that companies are selling them at a premium. Sometimes you have to sign a year contract just to guarantee the tools and the individual you want on location.

The influx of more tools and smaller companies will put horizontal drilling back in competition and lower the prices.

As of the writing of this book, it is still hard to get a fair price. The well needs to be planned out in advance in order to shop for good prices and get a good deal. Of course, wells that are promoted through phone rooms do not create a cost concern, since the cost is just passed on to the investor. That also is a factor in the high prices. Established oil companies that use most of their own money to drill the well have to operate more efficiently so the well will make a profit. One great advantage of the horizontal well is that an investor has a better chance of a return on his money because of the many producing zones encountered. A vertical chalk well can only deal with one fracture.

In choosing a directional drilling company, get some bids and shop around. Choose a directional driller that has had experience in curve and horizontal drilling and has been successful. A good directional driller can make or break your operation. Once a well goes in the horizontal direction, the consultant's main job is to keep the well from blowing up and to work the PWD system. The directional part is really out of his hands, but he should stay on top of what the driller is doing. If it looks as if something is wrong, ask the driller and call the office engineer and give a quick report. Since the business is booming, there are many fast-trained directional drillers in the field. As more people are trained and acquire experience, things will improve.

One thing to watch for is the directional driller using the brake. Directional drillers want to move their tools, but they do not have the insurance to do this. If they hurt someone, the consultant will also be dragged into the lawsuit for not watching his business. The driller from the rig is the only one allowed to touch the brake. The directional driller using the brake is one of the serious things to watch for, for the safety of the operation, since he is not familiar with your rig. Any good rig driller will tell you that it takes a couple of days to adjust to a new rig.

The directional driller should have his own phone, or else your phone will be tied up all day.

The MWD (measurements-while-drilling) equipment is the best invention made for directional wells. The equipment is very expensive and is run by batteries. By comparison, the steering tool comes from the Dark Ages of drilling. The equipment gives a reading so that the directional driller knows where he is in the well. The advantage is that no wires are needed to take a reading, unlike the method using steering tools. This equipment can run four to six days on a set of batteries. After the batteries go dead, a wireline can be dropped and attached on the top of the MWD and retrieved. Then it is sent back down the hole to operate another four to six days. It is a first-class operation. The only problem is that the drilling must be stopped to get a reading. However, this is not a long time. The MWD tool costs from $50,000 to $65,000. Be sure to sign up for tool protection on the rental agreement. This tool protection will pay about half of the cost.

Also sign up for tool protection on all downhole motors; this will save you one-half the retail cost.

23
Drilling the Curve

Drilling the curve is probably the most important part of drilling a horizontal well. Some directional drillers just lose it when it comes to the curve. I have heard horror stories about 1 to 2 weeks of drilling the curve, getting the curve drilled, and finding out you are 180° off. One problem arises from trying to drill the curve too aggressively, using up to 20° to 26° per 100 ft. It is hard to make a curve with that kind of aggressive drill pattern. The best wells use 16° to 18° per 100 ft of build. Try to plan the well where you do not become too aggressive. The more aggressive the build, the greater the chance of failure. Also if the well is too aggressively built, it is possible to have pipe failure, sticking pipe, and more drag in and out of the hole. Drag can create several problems in the bottom hole assembly. In some instances the motor was unscrewed and left in the hole. This, of course, is due to not thread-locking in most cases.

When the intermediate casing is set too low, then the curve will have to be more aggressive; when the casing is set higher, then the curve can be smoother, easier to drill, and easy to enter and leave.

Another problem encountered is that of taking a kick while drilling the curve (see Figure 23-1). A lot of kicks occur when the last part of the curve is made. Since you are in the target zone and you can hit a fracture and take a good kick, it is important to be alert to a kick at all times, but especially when horizontal drilling is begun. In the past, a couple of wells caught fire while the curve was being drilled, so a good consultant will have all the men in position and ready. The problem also is that you have curve-building tools in the hole, not horizontal ones, so you may be forced to kill the well just to come out and change tools. I never like to see heavy brine water put in a well, because, in my opinion, it ruins the formation and kills half the well. This, however, is not so critical when the curve is being made,

Figure 23-1. Taking a kick on a curve. (Courtesy of Davenport Horizontal Drilling Consultants, San Antonio, TX)

since the horizontal well has not been started. Many operators have stopped drilling operations and made a well after a good curve kick, but this is due to their inexperience in the chalk. If a well is allowed to go 3,000 ft or more (the longest on record at this writing is 6,000 ft), then the operator should try to go the distance, not get scared and quit, figuring a well has been made. The more fractures an operator has to work with, the more productive the well will be.

The directional driller keeps track of exactly where you are in the curve. If the well is not making the proper curve, then the directional driller might pull the tools and add a more aggressive bend tool, or he might add a kick pad to the motor to make it bend more. The kick pad has its pros and cons. A lot of kick pads have been lost in the curve and had to be fished out, but most of the time they will be buried in the curve cuttings or will not bother the tools. This call for a fishing job will come from the office engineer. Remember this: If the motor is hurt, then the operator will pay for it, so using a kick pad is still up in the air. My suggestion is to build a kick pad into the tools and use them only in special areas, where the curve is not doing what you want. Thus the pad will not get lost. If a kick pad is used, make sure it is welded securely. It is much easier to decrease the aggressiveness of the curve than to increase it.

Another point of interest: Have the directional driller start the curve, if possible, above the kickoff point, so you can add a margin of additional feet to make the target area. For example, if the kickoff point is 30 ft below the casing, then start the kickoff at 15 ft. That will give you an additional 15 ft to play with, and in some cases that is enough to make the curve a success. All the nice office plats and graphs look good on the wall, but using common sense can make it better in the field. The curve has so many variables that the directional driller may need assistance from more experienced hands, so do not allow the directional driller to keep coming in and out of the hole; if things are not right, ask him to get some help. If he will not call for help, then call the office engineer and explain what is going on, and he will call the service company and get some help on the way. Sometimes two heads are better than one. If a directional driller has a hard time on the curve, it does not mean he is incompetent; but he is not doing something right and needs help. It is not uncommon to see several hands out on location while the curve is drilled. Sit down with the directional hand when he arrives and discuss the program with him. Tell him you know the curve is hard and he may need help, and tell him not to try to bluff his way through at the operator's expense. Some of the sharpest hands I have met in the field knew how to use the phone and ask for assistance. Also explain to the hand your policies and procedures on location, and inform him that he has to attend all the safety meetings on kick. This applies to his assistant and the MWD hands, too. They are now part of the team, and a location can only have one boss, the consultant or field engineer. Good communication with the directional hand will make the job go more pleasantly and save money for the operator. And if the operator makes money, then drilling continues and all parties benefit.

Figure 23-2 is a simple drawing to explain how the curve is made. The drawing shows the depth of the casing and the kickoff point and the target zone. As the well is being drilled, the directional driller will plot his success, and you can find out if the curve is going as planned. A quick way to check the plat is to figure the depth needed to hit the target zone and divide that by 100; then divide 90 by this number, and you will get the necessary degrees to make the curve. For example, say it is 453 ft from the casing to the

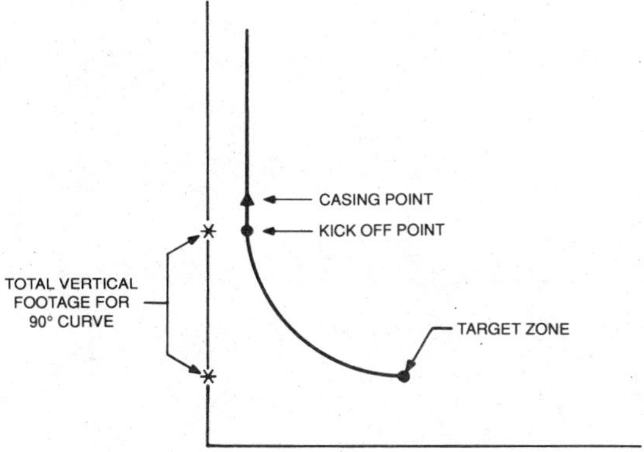

Figure 23-2. How a curve is made.

target zone. Give yourself about 10 ft to start under the casing, so that is 443 ft. Divide 443 by 100. That gives you 4.43. Now take 90° and divide it by 4.43, and you get 20.31°. This curve will have to average about 21° to hit target. If it is too deep, it is hard to come back up, so this well will be harder to drill.

Some operators who have encountered this problem have been able to head the pipe back up to the target zone, but it makes the movement of pipe more critical. If this happens, sometimes the operator will go back up the hole, cement the well, and start over again. This is very expensive and time-consuming. Another thing that is unique to this way of drilling is that after the horizontal well is depleted, you can pull up the hole, cement it, and then drill a new curve to a deeper zone below, using most of the old curve. I am sure that very soon horizontal wells will be drilled in all directions and take oil from all directions. The possibilities are exciting.

24
Special Problems During Horizontal Drilling

The problems encountered in drilling are so unbelievable to the normal drilling operation that many have a hard time believing consultants at the coffee table. This chapter discusses the most common problems. Expect the unexpected in horizontal drilling. These subjects are covered:

1. Drilling while taking a kick
2. Drilling with no returns
3. Drilling while on a vacuum
4. Leaking of the rotating head
5. Rotating head rubbers that leak and their removal
6. Producing too much oil
7. Taking a kick while tripping out of the hole
8. Spreading the kick
9. Drilling through the annular preventer
10. Sticking pipe coming out of the hole
11. Tripping out of the hole under pressure
12. Figuring kill weight brine water

13. Rain on location
14. Making sure all hands are trained
15. Locating enough frac tanks to handle a major kick

Drilling while taking a kick. When horizontal drilling began, the accepted way to drill a horizontal well was to drill into a fracture and let it flow; if it was apparent that it was going to produce more than the fluid capacity of the holding tanks, then the order was given to pump 10 lb brine water down the hole and kill the well. This went on quite often, and it ruined a lot of wells that could have produced more. It was a great rest period for the directional driller since most wells would flow for one to two days before being killed or depleted. The consultant would be up until the well slowed down. It was quite exciting to see the flares and the smoke, and the operators would start counting their money. But who really made the money was the directional company that was resting in its trailers. The day rate continues when a kick is in progress. All the back slapping and toasts and all the oil produced and sold just barely paid the directional company's fee. So now we are back to square one. Let's get the well drilled as quickly as possible. Now the operators are realizing that the longer the directional company stays on location, the lower the profits will be. Promoters used the flares and smoke to promote the next deal, and it worked. But soon the wells slowed down and the investors were not getting their money back as they had thought they would, so funds slowed down.

The horizontal well can make money if the well is drilled quickly. To do this, one must drill while taking kicks. This is done by adjusting the choke to hold a little back pressure on the well. If the pressure is too great, then close the annular preventers and slide the pipe if you can. The position of the kelly has a lot to do with drilling under pressure as well as the position of the directional motor. If the horizontal string needs to move up the formation, then the pipe may have to be rotated, and with the annular preventer closed, this may be impossible. If the kelly is up, then you can drill about 30 ft more; this gives the formation time to slow down. If the rotating head can take the pressure, you can rotate the pipe or slide. The key is how much pressure the rotating head can take. If the drilling operation is not stopped to produce oil, then the cost to drill the well will be lower. The oil will still be there, so do not think that it is now or never. When you are making up a joint, while the well is flowing, have the choke wide open and let the gas out of the flare line. This relieves the pressure on the rotating head. After the joint is added, close the choke to keep the well from blowing full blast. Normally leaving the choke 60% closed will keep everything in balance. But keep a sharp eye for necessary adjustments. A consultant should be on the floor when all this is going on, for safety purposes. When a connection is made, be sure the downhole back-pressure valve is working. If the well starts to flow from the drill pipe, then the valve has failed. Immediately reconnect the kelly to the drill pipe, and go back in the hole. Then determine the next step to take. This can be tricky because a joint needs to be added to keep drilling. One way is to break the kelly, if the flow is not too great. Stab a TIW valve in the open position and then close it; make up the kelly with a new joint and add it to the string with the TIW valve. After it is made up, open the TIW valve, go in the hole, and continue drilling. The TIW valve can be taken out after the well stops kicking. Just remember that it is there. Another way to add a joint is to pump a slug of heavy brine in the drill pipe; after it gets to the bit, break the kelly, and the heavy brine water will keep the formation pressure from coming up the drill pipe. This will allow you to keep drilling the well. As soon as the string needs to be pulled, the valve can be replaced. Remember when taking

Figure 24-1. Drilling while taking a kick. (Courtesy of Davenport Horizontal Drilling Consultants, San Antonio, TX)

a horizontal kick (see Figure 24-1), you can shut in the well at any time without fear of its blowing because of the reduced pressures you're dealing with. It is like a water faucet. If things get off track, just shut in the well and reevaluate your position and the best way to handle the problem.

Drilling with no returns. Drilling with no returns is common in some wells. The main thing is to keep the water coming, and in most cases the freshwater pump may not be able to keep up with the amount of water used to drill the well, so it may be necessary to pump out of the reserve pit. I have been on some wells where we got no returns for two or three days. We pumped everything on location down the hole and sometimes had to bring in field brine to keep up with the drilling operation. Field brine is available in most chalk fields, and instead of taking the brine water to disposable wells, it can be used in the drilling operation. If you run out of water for a short time, do not panic. Just quit drilling until you can build up enough water. A smart consultant will fill the reserve pit while the vertical part of the well is drilled to get ready for the horizontal part. You cannot have too much water on a location.

Drilling while on a vacuum. Drilling while on a vacuum is really something to see. It is caused by drilling into a fracture that has been depleted; it actually sucks air into the ground. You may drill on a vacuum for one or two days or maybe 6 hours. Do not worry; just keep pumping the water while drilling to help the bit drill better. Eventually it will fill up, and you will get returns.

Leaking of the rotating head. Leaking of the rotating head is a serious problem, especially when taking a kick. Normally the leak will be from the O-ring. The rotating head is vibrating all the time, and it is not uncommon for it to come loose and leak. If you are taking a kick, close the annular preventer and set the choke wide-open; then very carefully open the shale shaker valve to release any pressure under the rotating head rubber. Make sure, when you do this, that no one is running anything that could spark. The gas will sound like a freight train going to the shaker until the pressure is gone, so open the valve very slowly. Then with the pressure off the rubber, shut in the well and tighten the nuts that nipple the rotating head on the stack. Even though the well is shut in, take every precaution not to make any sparks under the floor. After the repair is made, go back to drilling. Also close the shale shaker valve.

Rotating head rubbers that leak and their removal. When rotating head rubbers start to leak, removing them can be tricky. It is important to understand that the rotating head rubber is the first defense against a well's flowing. The second defense is the annular preventers, and the third is the pipe rams. After the rubber is used for stripping in and out the hole, it starts to wear out. Watch for leaks all the time. If there is a leak during a kick, the annular preventers must be shut in and the rubber removed with the kelly. After the annular preventer is shut in, release the pressure between the annular preventer and the rotating head. This can be done by slowly opening the flow line valve and bleeding off the pressure. Then a lock ring on the rotating head needs to be released. While the locking device is released, make sure no one is looking over the kelly bushing, so that if pressure is released, it will not hurt anybody. After you make sure that no pressure is on the rotating head, pull it with the kelly, put a new rubber on it, and reinstall it in the head. Then after it is locked down in place, reopen the annular preventer—slowly so as not to hurt the new rubber. It is a good idea to set the choke wide-open to release as much pressure as possible below the annular preventer; then open the annular preventer and let the pressure hit the rotating head. After everything is OK, go back to drilling.

Producing too much oil. Producing too much oil (see Figure 24-2) is a problem if the operator and the consultant do not understand how to shut in a well while oil is being transported off location. In drilling a horizontal well, you can shut in the well at any time to wait for transports to remove the oil that has been produced. Many operators panic and order frac tanks by the dozen. It is a fiasco, and it costs money. Drilling may have to stop while the oil is being transported off location, to make room for more oil storage. Remember, every frac tank delivered to the location costs money, and if the well has three to four frac tanks, that should be enough to handle most kicks. There is always an exception to the rule, so a good consultant always knows where to get more tanks. It is also important to have lined up the tankers and the oil treaters to treat the oil before it is sold. The treaters have a flow system built into the choke system so some treating is going on as the well flows. But when pure crude comes down the pipeline to the gas buster, more treatment is necessary. Also the operator must be ready

Figure 24-2. Producing oil. (Courtesy of Davenport Horizontal Drilling Consultants, San Antonio, TX)

to file an emergency sale of the oil with the regulatory powers. All this needs to be coordinated before a kick. The basic thing to remember is this: The well can be shut in if you get in a bind, and it will give you time to move the oil.

Taking a kick while tripping out of the hole. Taking a kick while tripping out of the hole can be tricky. When the order is given to trip out of the hole, check for flow before moving the pipe. If the well appears to be dead, then trip pipe to the intermediate casing and check for flow. If there is no flow, then it is probably OK to trip out of the hole with no problems. It is very important to keep the hole full as you are tripping. Fresh water will hold most horizontal wells in place to trip pipe. Some operators will trip to the casing shoe then send and circulate a 10-lb slug of brine water and circulate it in the casing to the surface. This, however, has pros and cons. If this is done, before you go back to bottom, make sure that this heavy brine water is circulated out of the well, and then trip back to bottom. Heavy brine is detrimental to the success of horizontal drilling. If the well kicks while you are tripping pipe, determine where you are, shut in the well, and release the pressure with the choke system. Then reinstall the rotating head rubber, and open the annular preventer, checking for leaks around the rubber. If it is OK, then strip back to the casing shoe and circulate a slug of calculated kill weight, which you can determine by the shut-in pressure of the drill pipe. Pump the slug back to the surface, then check for flow. If there is no flow, slowly trip out of the hole. Normally this will work well. Instead of tripping back to the casing shoe, some operators will bullhead (force down under pressure) the fluid to the shoe by closing the annular preventer and choke and pumping an estimated number of barrels of heavy fluid down the hole. I personally do not like the bullhead method, since the horizontal wells are much easier to control with light fluid. After being in the horizontal part of the well, some operators trip out through the rotating head rubbers until they get to the surface. The rubbers are not very expensive, but if you make five or six trips, then it could become costly. The consultant should be able to make that call when it is time to trip.

Spreading the kick. Spreading the kick is a method developed by me for killing a kick with fresh water. It has been used very successfully on many wells. The idea is that not all fractures in the chalk have oil or gas; some are empty. In my experience, you take a good kick, then drill 30 ft more or so, and hit another fracture that is empty; and the pressure, instead of going up the hole, spreads the other way. This always stops the pressure from coming up the hole. The key to this system is to keep drilling. This is the time when the directional driller does not like working for you, because you keep him up around the clock and make him work instead of shutting down and letting the well flow. You need a rotating head rubber that can handle the pressure. If the rotating head rubber cannot hold the pressure, you have to close the annular preventer, grease the kelly with pipe dope, and, if you can slide the drill pipe down the hole, keep drilling. If the pipe needs to be rotated, then very slowly turn the kelly after greasing it well with pipe dope. This is not too dangerous as long as it is well greased. The kelly is eight-sided, so turning it will not hurt the rubber. The toolpusher will tell you it cannot be done, but it can. This is one of those items that needs to be discussed at the prespud meeting. In some cases the pipe must be rotated with the annular preventer closed. You apply only enough pressure to keep the kick pressure out, so it is not a full-pressure shut-in. If it starts to leak, which is highly unlikely, you can apply more pressure to the annular preventer rubber; this should seal it in most cases. But if that fails, the kelly will have to be raised and the pipe rams closed. Thus the drilling operation must cease until the pressure decreases to the point where drilling can continue through the rotating head rubber. Remember, in all these problems the well can still be shut in and the flow stopped, to allow oil to be transported off location. The above problem might occur rarely, but it could happen.

As a consultant, your job is to remain cool and calm, especially in front of your hands. If you can ''spread the kick,'' the well will slow down and you can make the repairs to the blowout equipment. The one or two joints you have to drill will not take that long, and once you hit another fracture, the kick will go the other way. The main thing is to not panic and start pumping heavy brine water down the hole, because then you mess up the well. Some will disagree, but I have seen too many good wells kicking, only to have heavy brine water pumped down the hole and never to see much oil afterward. The fractures are very delicate and heavy water appears to create cracks in the formation, letting the reserve move to another part of the formation. If you are drilling while taking a kick, adjust the choke to keep 100 to 200 lb of back pressure on the formation. This will also help control the flow to the gas buster. If the gas buster is overloaded, the gas will come out of the buster and into the skimmer tanks. This becomes dangerous if the derrickman working the tank creates a spark. The only way to control the gas buster is to use the choke to control the volume. Just keep in contact with your derrickman, and he will give you a status report. The hydraulic superchoke is the consultant's main responsibility on the well; without the choke, drilling a horizontal well would be virtually impossible—it would have to be done manually. Remember, once a good kick occurs, get to the next fracture as soon as you can, and you can spread the kick.

Drilling through the annular preventer. Drilling through the annular preventer is done by only one operator I know of, and he is enjoying great success. The Williams Tool Company has a 900-psi rotating head now, and it is normally all you need; but I have been told they also have a new one coming out that will hold 1,500 to 2,000 psi. When that head comes out, I am sure the operator who uses two annular preventers will go to the new system. If you get in a bind, the eight-sided kelly can be rotated in the annular preventer without much problem, as long as it has pipe dope on it.

Sticking pipe coming out of the hole. Sticking pipe coming out of the hole can be solved very simply, by adding some polymer. It can be added at the rotary table by pouring it down the drill pipe. Normally if the pipe is becoming difficult to pull, a couple of quarts will sweep the hole and that should make everything slick. Never force the horizontal portion of the string out of the hole; take your time. Usually you add 1 quart of polymer per joint of pipe to keep the pipe from sticking.

Tripping out of the hole under pressure. Tripping out of the hole under pressure is not hard, but it requires more attention by the driller and toolpusher and a safety meeting with the crew and the consultant. First, the well could be killed once in the intermediate pipe, but killing the well creates complications for the formation, if the heavy brine gets in the horizontal portion. Stripping to the casing shoe is done through the rotating head rubber. It is very simple, but watch out for leaks.

Figuring kill weight brine water. Figuring the kill weight of brine water is simple. Use standard blowout calculations, and remember to use only the true vertical depth when you calculate the kill weight.

Rain on location. Rain on location may seem like a silly subject. But if you have been in the Pearsall field and been hit by a rainstorm, then you know what this is all about. First build the location so that the water will drain off and the roads will not be under water if it rains. If you have a low area going to the location, then build a road with a culvert so that the water will drain off and not cover the road. Make sure that the water will not fill your waste hole around the trailers. You can build a dyke around the waste holes so that if it rains, they will not fill in and there will not be human waste floating all over. Make sure that the truck routes are higher or built so they will drain quickly, in case you are taking a kick and need to send trucks back and forth. It is very important that the consultant get the location ready for bad weather.

Making sure all hands are trained. Making sure that all the hands are trained is a full-time job. I have never been on a location where the toolpusher does not know how to figure the kill weight. The boom in the oilfield has created a new generation of roughnecks and young men entering a new career. Most do not have a college degree, but they can learn very quickly what is going on, if you spend time with them at safety meetings. Most consultants working on land rigs cannot figure the kill weight mud if asked. I believe that all consultants should be required to attend a blowout school once a year. I require all mine to be blowout-certified, or else they cannot even apply for a job. If more operators required this, the older, toolpusher-type consultants would be forced to go to school or go back to roughnecking, and I mean roughnecking not drilling or toolpushing, but not in a capacity of making decisions. It is simply too dangerous to work in the chalk if you do not understand downhole pressures. One of the best schools around is Well Control School based out of Lafayette, Louisiana. The school itself is in Houston, Texas, and is taught by one of the best, Mr. Dee Avery. I am sure that soon all supervisors will be certified. The best way to teach the crews is to get a chalkboard and hold a 30-minute meeting with each shift until everyone can figure the kill weight mud. When I take a kick, sometimes I shut it in so the hands can practice what I have taught them. This also makes them more confident when trouble occurs. The hands who work on my wells leave the job with a working knowledge of well control.

Locating enough frac tanks to handle a major kick. Locating enough frac tanks to handle a major well kick can sometimes be a problem. If you understand what has been written in this chapter, then you know that covering the location is not necessary. But every once in a while you have to work for someone who tells the engineer to order

Figure 24-3. Bringing in frac tanks. (Courtesy of Davenport Horizontal Drilling Consultants, San Antonio, TX)

out every frac tank in the world (see Figure 24-3). Sometimes if you want to keep your job and get the next, you have to follow an idiot's program. When the location is built, be sure it is big enough to add more frac tanks and handle the trucks on location.

25
Setting the Packer in a Horizontal Well

After the well has reached the horizontal TD (total depth), it is time to come out and set the packer. Make sure the well is not flowing when the order is given to come out of the hole. If it is flowing, you will have to drill to the next fracture or trip to the casing and pump a slug in the inter-mediate casing and check for flow. Once the flow is stopped, then trip out the hole. Now comes a small problem. Make sure the wireline company has the release tool for the packer you are running. If it does not, call the packer company and ask them to deliver the packer with the release tool. As hard as it is to believe, most packer companies have packers but no release tool; it has to be ordered in advance. Once you come out of the hole, with the bottom hole assembly, it is open to the world until you get the packer in place. However, you can close the blind rams and strip back into the hole, using the rotating head rubber, after you open the blind rams. In case of problems, make sure the packer hand is on location and almost rigged up when you come out of the hole. You need to set the packer on the last 40 or 50 ft of the bottom of the casing. If you use a retrievable packer, then you can always pull it when the well is depleted; you can use it again after repacking it with new rubbers.

After the packer is set, the well is finished for the consultant. So the only thing left is to make sure the packer is not leaking and to start nippling down the rig. After the blowout preventers are taken off, you may put a plate over the well and bolt it down. The blind flange will keep the well intact until a completion crew arrives to run tubing and complete the well. While running the packer, make sure you use a lubricator in case the well starts flowing while you are setting the packer.

26
The Long String and the Cement Job

The order to run the long string comes after careful evaluation of the logs, core samples, cuttings, and DSTs. The geologist and the engineers at the office decide what grade of pipe to run and whether a DV (stage) tool is going to be run to cement around production zones higher up.

Usually the pipe is already on location, strapped, numbered, and ready to run in the hole. If not, it is usually on the way. When it does arrive, as always, the consultant should personally strap the pipe, and lay it out in rows, ready to run, with the collars facing the V-door. After the first row is delivered, the pipe is numbered from the end, working toward the V-door. The casing should be strapped and recorded in rows of ten on paper to simplify tallying the pipe.

The truckers will have to wait while the pipe is strapped, row-by-row. This will probably upset the trucker, but the pipe is more important. Even if the trucking company charges for waiting time, it is more important to take time and eliminate mistakes on the pipe tally. (For some reason, mistakes in tallying are rampant in the oil patch. If you take the time to tally the pipe accurately, you will be a hero with the operator.)

Example

The hole is +9,200 ft and you received 221 joints on location. You must figure the pipe as close to 9,200 ft as possible. Assume there are 219 joints totaling 9,240 ft and two short joints, one 30 ft and one 36 ft for a total of 9,306 ft.

Joint 47 is 41.72 ft, joint 56 is 42.10 ft. Pull them plus the 30-ft joint, leaving 9,192.18 ft for the long string. Place the other short joint, three joints above the float collar to serve as a locator when completing the well. Tag the pulled joints with red tape so that they will not be run into the hole. Now add 1 ft for the guide shoe and 4.71 ft for the float collar, bringing the string up to 9,197.9 ft. This is about as close to 9,200 ft as you can get.

Since the hole has been drilled to 9,200 ft, we will have a rat hole of 4 or 5 ft, which is acceptable. Having a rat hole does not harm the production string. If the pipe had been on location before TD, you could have simply told the driller to go to 9,197.9 ft. Since you had to order out the pipe and did not have a tally, you had to drill to the TD called for in the prognosis and adjust the string by pulling joints out.

Pipe is delivered in advance to save rig time and money in case the decision is made to run the string, but if the string is not run it is fairly expensive to move the pipe off location. It is debatable as to which way saves more money. If time is a factor, it is best to have the pipe on location.

Condition the hole and the mud for running the casing while the pipe is being shipped. Have the mud engineer come out and make sure the mud is ready. Also make arrangements to sell the liquid mud if at all possible, since after the hole is finished, the mud is no longer needed.

When the hole is conditioned, the job is almost finished. The next step will be laying down the drill pipe while tripping out. Order out a laydown machine and three or four hands to roll and stack the drill pipe as it comes out of the hole. The operation takes 6 to 12 hours, depending on the hands and the depth.

When you begin conditioning the hole, order out the casers, the laydown machine, and the cement crews. Normally the casers and the laydown machine are furnished by one service.

To condition the hole, circulate bottoms-up once. (This means to bring the mud on the bottom to the top. Bottoms-up can be determined very easily if you know the bit-to-surface time. The mud logger can easily provide this number.) After circulating bottoms-up once, make a 15-stand short trip, then circulate bottoms-up once more.

When the laydown machine arrives, have the laydown crew rig up their unit. When they are ready to go, give the order to pull off bottom and chain out while laying down pipe. The operation takes about 1 to 1.5 hours per thousand feet of pipe.

When coming out of the hole, chain out to keep from causing fill to fall down the wellbore. Fill can cause problems if too much falls to the bottom. Once the bit is in the casing, the rotary can be used to break out pipe.

Scheduling the services is important when running pipe so you can eliminate wait time. The following list should provide a general guide for scheduling the services needed.

1. Circulate and short trip time—approximately 6 hours.
2. Rig up laydown machine—1 hour.
3. Trip out, laying down pipe—1 to 1.5 hours per 1,000 ft (10,000 ft—15 hours).

4. Rig down laydown machine, and rig up casers—1.5 hours.
5. Install casing rams in BOP in case of a blowout—1 hour.
6. Run casing—10 hours.
7. Rig down casers—1 hour.
8. Rig up cement crew, break circulation on casing, circulate bottoms-up—1.5 hours.

Normally the laydown machine and the caser crew will come out together. The casers will roll and stack pipe. Order them out when you start circulating and conditioning the hole. By the time you are ready for them they should be on location.

Next order out the cement crews. Look at the time schedule to have them out just before the casing is run. Using the preceding guidelines, you should have them arrive 23 to 24 hours from the time you start circulating and conditioning the hole. Remember, the crews charge by the hour and every hour they are on location costs money.

While the casing is being run, make sure the rubber thread protectors are used so that the pipe threads will not be damaged. Make sure that any bad thread protectors are not used and returned for credit. This is important as the thread protectors are very expensive.

Go over the cement figures with the cement engineer. Most long strings have two mixtures of cement, one to lead and one to tail. It is the consultant's responsibility to ensure that the right combination leads and the right one tails.

Example

The cement will be run from 9,200 ft to 6,500 ft for a total of 2,700 ft. Use the cement book to find the following information:

1. Volume and height between 7-in. 23 lb/ft casing in
   9⅞-in. hole. Find the annular volume:

   2,700 ft $\times$ 0.2647 ft^3/ft
   $$= 714.69 \text{ ft}^3 \text{ (715 ft}^3 \text{ rounded off)}$$

2. Displacement of cement with mud down to float collar
   which is 47 ft above shoe.

   9,200 ft $-$ 47 ft $=$ 9.153 ft

   From the cement book under capacity of casing, find
   barrels per foot:

   9,153 ft $\times$ 0.0394 bbl/ft $=$ 360.62 bbl (rounded
   to 361 barrels)
   needed to displace
   the cement to float
   collar

3. Figure cement at 20% excess to cover washouts, etc.:

   715 ft^3 $\times$ 0.20 $=$ 143 ft^3
   715 ft^3 $+$ 143 ft^3 $=$ 858 ft^3 of cement

4. The lead cement will be 450 ft^3 at 1.19 yield per sack,
   so we get:

   $$\frac{450 \text{ ft}^3}{1.19 \text{ yield}} = 378.15 \text{ sacks (rounded to 378 sacks)}$$

   Converting to barrels:

   450 ft^3 $\times$ 0.1781 $=$ 80.14 bbl (rounded to 80 bbl)

5. The tail cement is 408 ft³ at 1.20 yield. We get:

$$\frac{408 \text{ ft}^3}{1.20} = 340 \text{ sacks}$$

Converted to barrels:

$$408 \text{ ft}^3 \times 0.1781 = 72.66 \text{ bbl (rounded to 73 bbl)}$$

These calculations tell you to lead with 80 barrels, tail with 73 barrels, and displace with 361 barrels. Since you are finishing the hole, some companies will want to displace the cement with salt water or diesel instead of mud. The mud may be sold for credit. Check with the prognosis or the operator's engineer.

Install your casing rams to ensure a safe operation if the well kicks. Casing rams can be rented from most rental tool companies.

When the casing is run, remember to put the centralizers and the float equipment in the right place and use liquid metal lock to keep the float equipment in place. Always check the prognosis for proper placement of the centralizers.

When the casing reaches bottom, tag the bottom, then pull back one foot to leave room for cement flow. If there is a rat hole, just let the casing hang at the point where it is easy to handle on the floor in regard to the installation of the cement manifold. Set the slips on the floor and let the casing hang while the cement is being pumped. Make up the cement manifold and break circulation. Circulate bottoms-up to bring to the surface any gas or fluid that may have invaded the wellbore. After circulating the bottoms-up, the well is ready for the cement. (See Figure 26-1.)

Some engineers use two plugs. The first cleans the mud out of the casing and is followed by the cement. When the plug hits the float collar, it breaks down and allows the

Figure 26-1. Casing and cement trucks. (Courtesy of Davenport Horizontal Drilling Consultants, San Antonio, TX)

cement to be pumped down through the float collar. The second plug is installed and pumped after the cement to clean the casing wall of cement and to isolate the mud from the cement. If the casing is rusty, use only one plug because the first plug will collect rust and plug up the float collar, causing problems. It is better to use only one plug, though some people will argue for two. Try to talk the engineer into using one plug, since one is safer and less trouble.

Pump the lead cement and then the tail cement. Install the plug and displace the cement to the float collar with 361 bbl of displacement fluid. When the plug reaches the float collar, a noticeable increase in pump pressure will occur. This means the plug has been bumped and is on top of the float collar.

Check for backflow. If there is backflow, repump the regained fluid and close the cement manifold to hold the cement in place. (See Figure 26-2.)

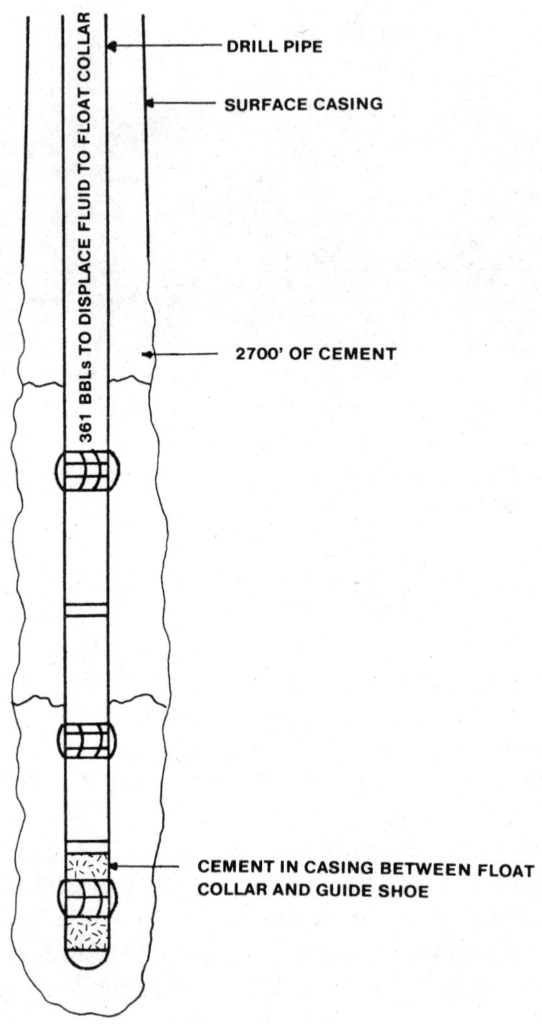

Figure 26-2. Cementing the long string.

Figure 26-3. Long string cemented. (Courtesy of Davenport Horizontal Drilling Consultants, San Antonio, TX)

Wait 12 to 18 hours for the cement to set. Remember to take four samples while the cement is being pumped to determine if the cement has set and how fast. After the cement job (Figure 26-3), release the cement trucks and crew.

On any long string the cement engineer and salesperson can be of great assistance. The cementer can run the job smoothly if given the chance. Let him run the job while you watch for mistakes.

27
Finishing the Well and the Paperwork

After the casing is run and the cement is pumped into place, the drilling consultant's job is almost finished. His next operation is to set the slips on the long strings. To do this, the BOP's will need to be nippled down from the casing head. Pull the BOP's up enough to enable the slips to open and to be placed around the casing. Then rebolt the casing slips and let them fall into the casing head.

Pull on the casing 50,000 to 100,000 lb, and then hammer the slips around the casing with a rubber hammer to set them into place. Slack off on the casing to set and seal the slips.

Get the welder to cut the casing one foot above the slips and to remove the cut piece of casing from the rotary table. Remove and lay down the BOPs and annular preventer (this is called nippling down the stack).

At the same time, have the derrickman start jetting the mud tanks. It will take about 8 to 12 hours to do this. When the stack is nippled down and the mud tanks are cleaned, release the rig and sign the daily driller's log, stating the time and date the rig was released.

You and the toolpusher should make sure all rental tools are found and are ready to be shipped off location. Call the rental suppliers to come pick up their equipment. If the suppliers have no one to send for the pick-up, have an 18-wheeler pick up 8 or 9 loads and deliver them to the suppliers. This will save much on shipping charges. An 18-wheeler can carry a lot of equipment and save time on loads going to one place. Be sure to get a copy of the pick-up invoice of all tools leaving the location to cover the delivery slips.

The last pick-up should be the consultant's trailer and the communication equipment. You never know when you might need to call for trucks or make emergency calls, since rigging down is always dangerous for the hands. Communication should be the last thing to leave the rig site.

Tell the operator the release time and approximately how long you will remain on location to remove rental tools. Double check each item to ensure that all equipment has left before you call in a finished job. Most consultants use a file box on location to help keep the paperwork in order. The box usually contains an A–Z folder, various folders with tabs, and office supplies and can be purchased at any office supply store. At the end of the job, all the paperwork should be placed in the file box and the whole file turned in to the company.

The paperwork is important to any operation and most of it, including invoices for completed work, parts ordered, and rental tools used, must be kept on location. It is best to keep these invoices in an A–Z folder according to date. It is very important not to misplace these invoices.

Receiving slips for all rental tools delivered to location will have a trucking bill. Keep all trucking bills for tool accountability.

The daily rig reports and the daily tour sheets need to be filed each day, preferably in a folder that is equipped with tabs at the top of each page. Usually two copies, the pink and the yellow, are kept. The bit records need to be turned in to help the operator or anyone else who may drill in the area again. The mud logger's report must be turned in with the file box at the completion of the well. Usually the loggers turn in their own reports, saving the consultant time and trouble.

Typing all the daily reports and having all your paperwork in order makes you look professional and will impress the engineer inspecting the location.

The following reports are turned in at the end of the job:

- Consultant's daily reports
- Daily mud reports
- Geolograph reports
- Pit indicator and flow show reports
- Invoices for services
- Rental tool receiving slips
- Rental tool pick-up slips
- Daily rig reports
- Bit reports
- Mud logger's report

Daily reports cover the day-to-day operation of the rig. They include:

- Well number and lease number
- Operator's name
- Daily mud evaluation
- Drilling costs, daily and accumulated, in detail
- BHA and length
- Daily geology report
- Footage of well and daily footage

The daily drilling report is a good way to look back at the well and review what happened at each depth.

Daily mud reports give all the mud data, including daily and accumulative costs.

Geolograph reports give a record of the weight on the string, downtime, and footage per 12-hour period. These records are kept in a book on the rig floor, and the sheets are collected and turned in later.

Pit indicator and flow shows reports are the record generated by the pit indicator and flow show machine. Simply roll the tape up and include it with the file box.

Appendix D contains several forms that will help you in your record keeping.

28
Plug-and-Abandon Procedures

If the well turns out to be a "duster" or dry hole or if there are no hydrocarbons present in commercial quantities, the operator will call the consultant and order the well plugged and abandoned. It will be necessary for the consultant to contact the state or federal agency that issued the drilling permit and obtain plug-and-abandon (P & A) instructions.

Before you plug the well, some of the intermediate casing may be cut and retrieved. Use a mechanical cutter to cut the casing five feet inside the surface casing. This will save the operator a great deal of money. Have a casing company come out with a crew to pull casing out of the hole and lay it down.

All government agencies require cement plugs above porous and high-pressure zones. They generally require cement *100 ft out and 100 ft in* the surface casing. This means you must put a cement plug 100 ft below the surface shoe and a plug 100 ft inside the surface casing. If an intermediate string has been run, the 100 ft out and 100 ft in also applies to this string. A 100-ft plug may be added near the surface, and an additional 10-sack plug may be set at the surface after the bradenhead is cut off below the surface. The surface plug is filled over with dirt. (See Figure 28-1.)

Example

When the order comes to plug and abandon, call the cement company and determine how much cement will be needed. You will need to obtain capacity information from the cement book. For a 9⅞-in. hole the cubic feet per foot is 0.5319. To find the cement needed for a 100-ft plug, multiply the cubic feet per foot by 100:

$$0.5319 \text{ ft}^3/\text{ft} \times 100 \text{ ft} = 53.19 \text{ ft}^3$$

To figure 20% excess:

$$53.19 \text{ ft}^3 \times .20 = 10.63 \text{ ft}^3$$

$$10.63 \text{ ft}^3 + 53.19 \text{ ft}^3 = 63.82 \text{ ft}^3 \text{ needed for each plug}$$

In this example four 100-ft plugs are needed so:

$$63.82 \text{ ft}^3 \times 4 = 255.28 \text{ ft}^3$$

If the slurry yield is 1.20 ft³/sack, to find the number of sacks:

$$\frac{255.28 \text{ ft}^3}{1.20 \text{ ft}^3/\text{sack}} = 212.73 \text{ sacks (213 sacks rounded off)}$$

Tell the toolpusher to strap in the hole with drill pipe, open-ended, and put the pipe at the depth of the first plug.

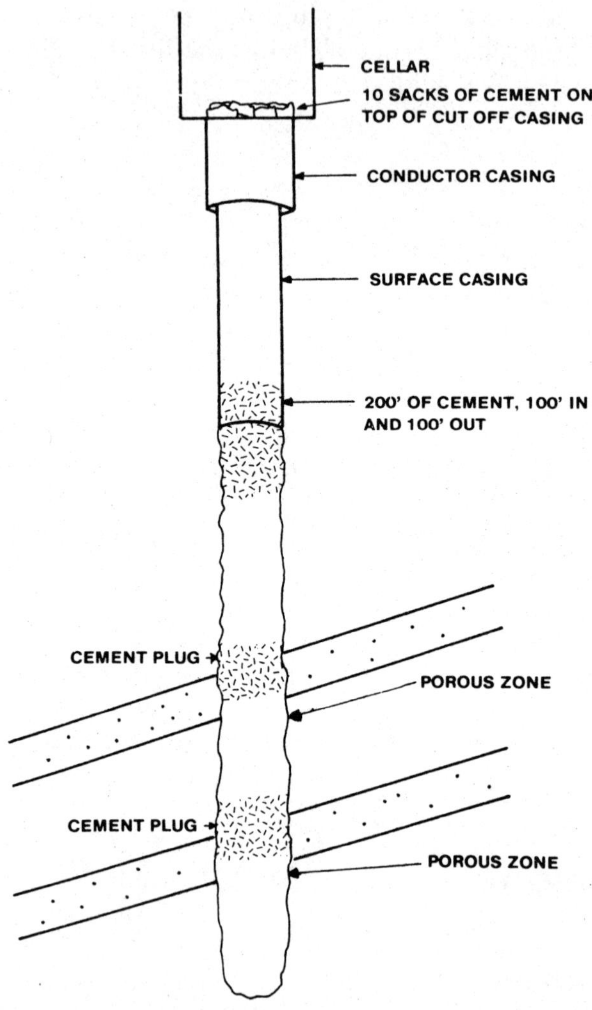

Figure 28-1. Plugging the well.

The drill pipe is 4.5 in. 16.60 lb, and the first depth is 8,900 ft. The cement book gives the capacity of this drill pipe as 0.01422 bbl/ft. To get the total amount of displacement fluid needed:

8,900 ft × 0.01422 bbl/ft
$$= 126.55 \text{ bbl } (127 \text{ bbl rounded off})$$

When the drill pipe is in position, pump 63.82 cubic feet × 0.1781, which equals 11.36 bbl, or rounded to 11.5 bbl of cement. Displace it with 127 bbl of displacement fluid. After the 127 bbl are pumped, break off the cement manifold and pull the drill pipe to the next plug site. Make sure the drill pipe is positioned at the bottom of the area to be plugged because the cement will flow up the annular portion of the DP, not down. When the DP is pulled, the cement stays in place.

When the drill pipe is in position, recalculate the displacement, reconnect the manifold, and circulate through the drill pipe. This will clean out any cement that may be left. Pump 11.5 bbl of cement and displace it with the new displacement figure.

Repeat the same procedure up the hole. Where 100 ft in and 100 ft out is required, pump 23 bbl (11.5 × 2 = 23 bbl) and position the bottom of the drill pipe 100 ft below the casing shoe. Recalculate the displacement and repeat the above procedure.

Remember, while setting the cement plugs downhole, you are also laying down drill pipe, so the laydown machine must be set up.

Some states permit setting a retainer plug in the casing. Fluid can be pumped down through it, but the plug blocks pressure coming back up the casing. This allows you to pump the pit and finish with cement, thus plugging the well (see Figure 28-2).

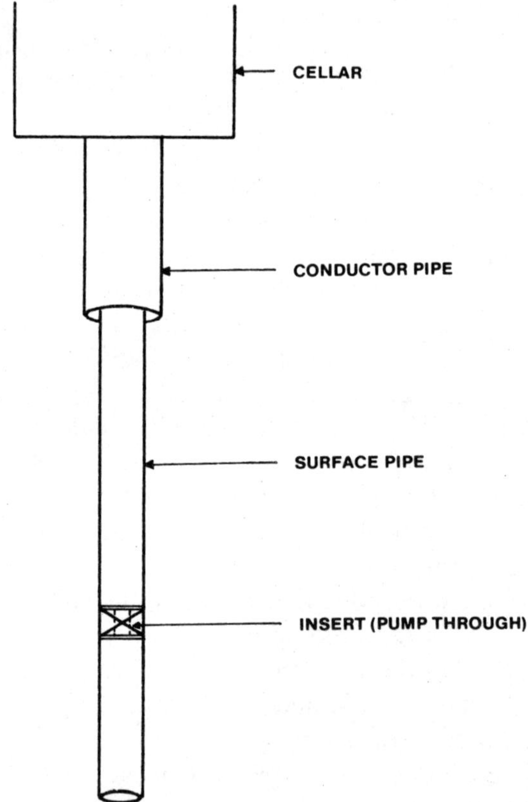

Figure 28-2. Insert in casing.

29
Blowout Control and Calculations

In the early days of drilling wells, a gusher was a welcome sight. Earthen dams were built around the well so the oil could be reclaimed and sold. This was before deeper holes and high pressure gas, which increase the risk of fire. The drillers drilled into a formation with the rig until there was either a blowout or a gusher. Today a gusher *is* a blowout and is definitely not a welcome sight. An army of state and federal environmental protection agents would be all over the rig to assess fines on the oil company, and in general, to give them a hard time. An offshore blowout is worse, since it also pollutes the water and attracts considerable adverse publicity.

Well control and blowout prevention have been developed over the years in the oilfield. Yet, no one has *all* the right answers, because blowouts and oilfield fires still occur.

An oilwell fire is the most dangerous aspect of drilling a well—and it happens all the time. An oilwell fire is simply a blowout that catches on fire. It can mean death, heavy financial losses, and bad publicity.

In this chapter the control of kicks and the associated calculations will be covered. This text is by no means the last word on blowout control, and perhaps someday the drilling business will be safer through the elimination of blowouts. But for now, our present technology will have to suffice.

The lack of skilled crews is the main cause of blowouts worldwide. The petroleum industry is growing so fast that many companies do not have the time to properly train their personnel on how to control kicks, which, if not properly handled, can result in a blowout. If you ask hands on a rig how to figure kill mud, they will shake their heads. They simply have not been taught. As hard as that statement is to believe, it is true! This is like driving a diesel truck down the road without knowing how to drive a car. Someone is going to get hurt.

Since blowouts are so common in the oilfield, the U.S. government has assigned the Minerals Management Service the responsibility of setting standards for blowout control schools around the country. These schools have helped tremendously in making the drilling business safer.

This chapter will cover the many methods used to control kicks, but it is advisable to attend a blowout school for more training and receive a certificate for supervisory work. The standards will change over the years, so you need to keep up-to-date on new procedures.

To simplify the business of well control, let us first examine the word kick. A *kick* is the entering into the wellbore of water, gas, oil, or other fluids associated with the formation. This occurs when the column of drilling fluid is lighter than the formation pressure and fluid or gas enters the wellbore (see Figure 29-1).

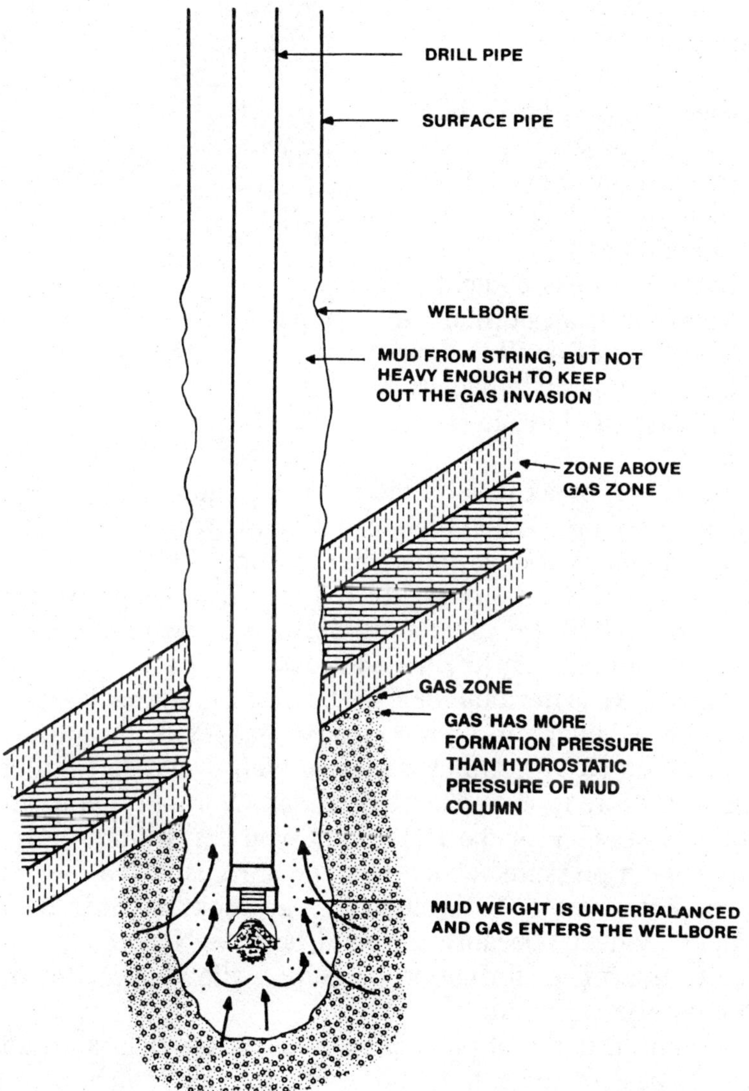

Figure 29-1. Drilling into a gas zone and getting a "kick."

If a kick is not controlled, a blowout may occur so it is necessary to notice the signs of a kick. They are:

- Rate of penetration increases
- Change in shape and size of cuttings
- Increase in rotary torque
- Increase in drag
- Sloughing shale
- Increase in gas content
- Variation from normal ''d'' exponent
- Increase in flow-line temperature
- Decrease in shale density
- Increase in chloride content

A drilling break usually indicates entry into a higher formation pressure. When higher formation pressure is hit, the mud weight becomes underbalanced and the drilling rate increases, sometimes dramatically, sometimes slightly. The mud logger will be calculating the rate of penetration and should notice the difference and report it.

Always keep the mud loggers and the drilling shack rigged up with an intercom to report any changes in the rate of penetration (ROP). Quick response with everyone notified can give more time to check for a kick. Normally, you will notice a slowing of the ROP while you are drilling on the top of high pressure, which in some areas is called the cap rock. After you drill through the cap rock, a faster ROP will be noticed, because the mud becomes underbalanced due to the higher formation pressures, allowing gas or oil to enter the wellbore.

A gain in the mud pits always indicates a kick is on the way or has entered the wellbore. The pit gain indicator will signal the driller on the floor when a pit gain is encountered. Have the derrickman record the barrels gained on the worksheet. The sooner the driller catches a pit gain, the quicker the kick can be controlled. After a pit gain is noted, pull the kelly up and pull out the bushing until drill pipe is in the annular preventer and then shut the annular preventer. Shut the pump off and check for flow. A flow indicates formation fluid or gas entering the wellbore and up the annulus, pushing the mud up the hole. Gas expands as it rises up the bore, and at the flow line it looks like the pump is on (see Figure 29-2). Shut the well in and start kill operations.

To shut the well in, do the following:

1. Raise the kelly to clear the drill pipe safety valve above the rotary table.
2. Shut off the mud pump.
3. Check for flow.
4. Open the choke manifold or hydraulic closing ram (HCR). This will avoid shocking the well.
5. Close the annular preventer. If the pipe rams are to be used, check that the tool joint is out of the rams.
6. Close the adjustable choke and watch the casing pressure.
7. Read and record shut-in drill pipe pressure (SIDPP), shut-in casing pressure (SICP), and pit volume increase (in barrels).
8. Watch and record the increase in pressure until the well stabilizes.
9. Check for leaks around the rig and the choke manifold and the BOPs.
10. Check the mud weights in the pits.
11. Make sure everyone is at the kick stations.

Shows of Gas, Oil, or Salt Water

A show will first be seen in the shale shaker area. It may be a small column of oil, which is easily recognized, or gas

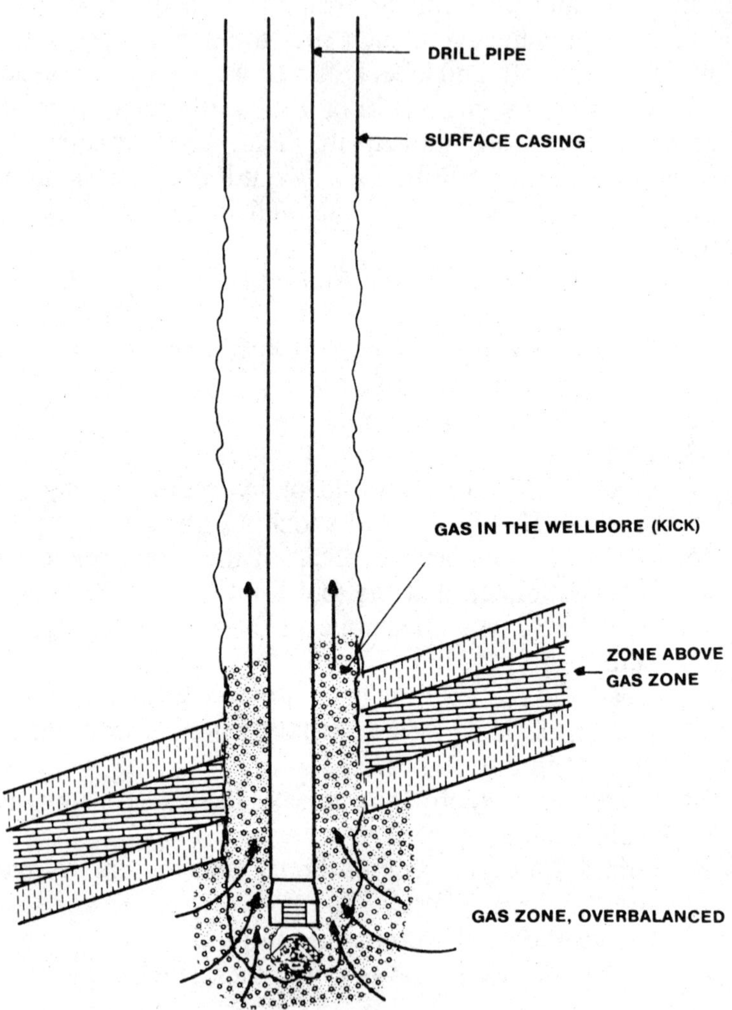

Figure 29-2. A kick.

cut mud. Salt water can be recognized by an increase in the chlorides in the system. When drilling in sands, make sure to hold back on the weight of the bit so you can watch for kicks and can handle the drilling better.

If gas enters the wellbore, the circulating pressure will decrease because of the loss of pressure balance in the annulus. Normally, the pump speed increases because of the difference in pressure and is easily recognized.

As a consultant on location, it is up to you to train the drillers to handle the paperwork when you are not on location. Always display and leave the information on the driller's wall, since in times of trouble people panic and do not think clearly. Train the driller and his hands on the procedures so everyone will be confident when trouble arises.

Ensure that all hands know their stations when fighting a kick. First, the derrickman and the youngest floorhand need to work at the mud hopper. The driller will work the pipe and handle the blowout preventers. The toolpusher needs to be available to work the superchoke and check for leaks around the rig. The motorman needs to be on the ground to check for leaks on the stack. In the dark, he needs to have a flashlight ready at all times. The floor hand needs to be on the mud pits to record loss or gain in the pits. The consultant needs a high-intensity beam flashlight *with fresh batteries* ready at all times. The rig also needs five waterproof flashlights. Make sure the pusher has these and shows them to you every two or three days in case they are needed.

If the hands know their places, fighting a kick becomes less confusing and difficult. Most people have been killed or injured because they did not know what to do or panicked in the face of unknown danger. A successful consultant will train the men on each crew every week on what to do and explain the nature of a kick to all personnel.

After the pressures have been recorded and everyone notified, the next step is to determine which method to use to kill the kick. There are three methods in the oil patch:

1. Driller's method
2. Wait-and-weight method
3. Circulate-and-weight method

Driller's Method

The driller's method is normally used on land rigs. It is an old method of controlling a well. It is not recommended by all U.K. blowout schools and is not taught, but material on the method is included in the handbook at blowout schools. The method is widely used and taught in Canada.

The method involves circulating the kick out of the hole, then a second and third circulation of kill weight mud. It is used on drilling rigs where crews are shorthanded and mixing facilities are slow. The only problem is the higher casing pressures. The method is simple, and it is easy for one or two men to do. The procedures are as follows:

1. Shut the well in after a kick is recognized.
2. Record the shut-in drill pipe and shut-in casing pressures.
3. Circulate the kick out of the hole.
4. Shut the well in a second time to build the mud weight.
5. Circulate the well the second time with the heavier mud.

Wait-and-Weight Method

The wait-and-weight is widely used in hard rock areas and overseas. On the Gulf Coast and in some sandy areas, it can, in some cases, get you in trouble by causing lost circulation and formation breakdown. This will be explained later. Wait-and-weight requires only *one circulation* to control the kick, which saves time and money. In hard rock areas, it is the best method to use where pore pressures and formation breakdown pressures are greater (see Figure 29-3). After you decide what kill weight is needed, the barite is added to the existing mud until the kill weight is achieved. Then the mud is pumped downhole to control the formation pressure. After determining the following information, weight up the mud and pump it down the hole. First, record all data on the well killing worksheet (see Figure 29-4).

Example

On the chart (Figure 29-4) we see the following:

$$
\begin{aligned}
\text{Well depth} &= 9{,}000 \text{ ft} \\
\text{Bit size} &= 9\tfrac{7}{8} \text{ in.} \\
\text{Drill pipe} &= 4.5 \text{ in. XO } 16.6 \text{ lb/ft at } 8{,}450 \text{ ft} \\
\text{Drill collars} &= 7 \text{ in. } 550 \text{ ft long} \\
\text{Casing} &= 10\tfrac{3}{4} \text{ in. } 40.5 \text{ lb/ft } 3{,}000 \text{ ft} \\
\text{Mud weight} &= 10.5 \text{ ppg} \\
\text{Pump pressure} &= \text{drilling } 2{,}600 \text{ psi at } 64 \text{ strokes per minute reduced circulating pressure is } 1{,}000 \text{ psi at } 30 \text{ strokes per minute} \\
\text{Shut-in drill pipe pressure} &= 300 \text{ psi} \\
\text{Shut-in casing pressure} &= 450 \text{ psi} \\
\text{Pit gain} &= 20 \text{ bbl (gas invasion)} \\
\text{Pump displacement} &= 233 \text{ gpm or } 5.55 \text{ bbl per minute at } 30 \text{ strokes per minute} \\
\text{Barrels per stroke,} & \\
\text{pump number 1} &= 0.185 \text{ bbl per stroke}
\end{aligned}
$$

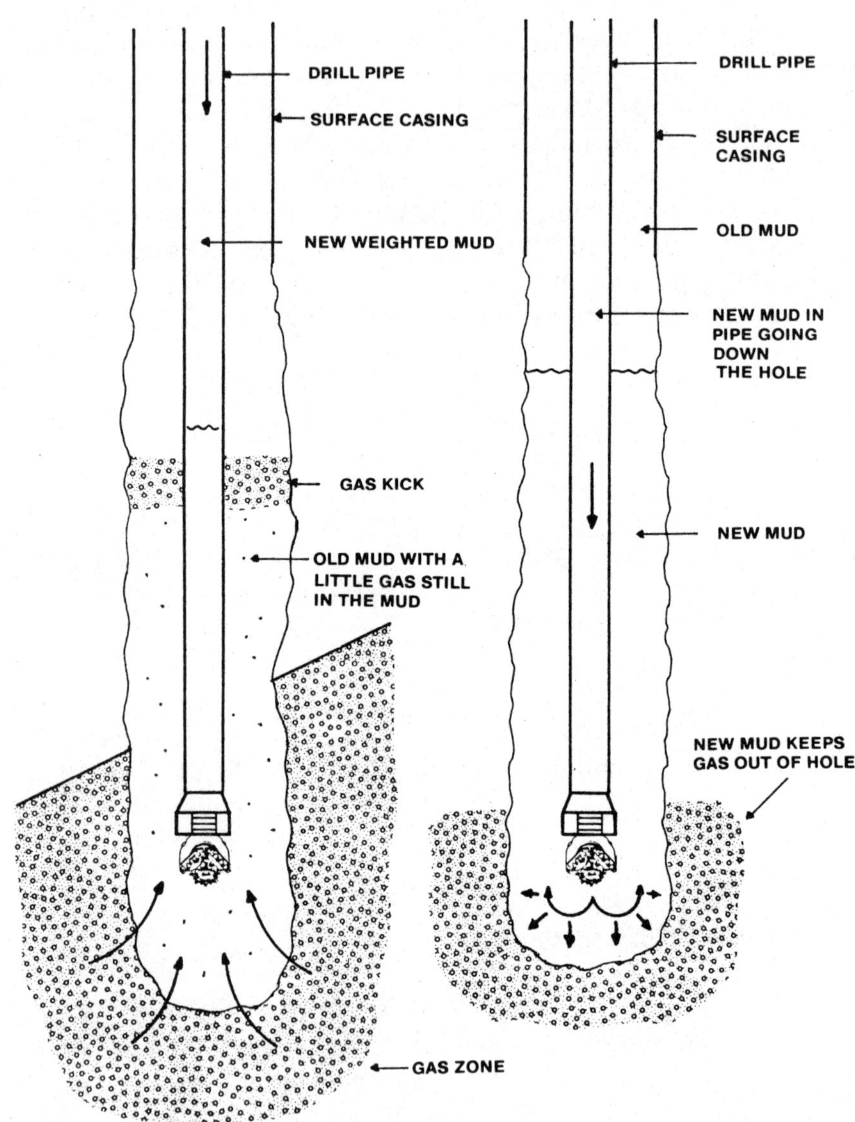

Figure 29-3. The wait-and-weight method for controlling kicks.

KILL SHEET DATA

PRE-RECORDED DATA

Original Mud Weight   =   _______________ ppg
Measured Depth   =   _______________ feet
Pump #1 SPP Rate   =   _______________ psi at _________ spm
Pump #2 SPP Rate   =   _______________ psi at _________ spm
Annulus Volume   =   _______________ barrels
Drill String Volume   =   _______________ barrels
Pump Output   =   _______________ barrels/strokes

Drill String Strokes   = $\dfrac{\text{Drill String Volume (\quad) barrels}}{\text{Pump Output (\quad) Barrels/Stroke}}$

                        =   _______________ strokes

KICK DATA

SIDPP   =   _______________ psi
SICP   =   _______________ psi
Pit Gain   =   _______________ barrels
True Vertical Depth   =   _______________

KILL MUD DATA

Kill Mud Weight   = $\dfrac{\text{SIDPP}}{\text{(TVD) x (0.052 feet)}}$ + ppg original mud

                        =   (\quad) ppg kill mud weight

PUMP PRESSURE

Initial Drill Pipe   =   SIDPP _________ psi + SPP _________ psi
       Pressure   =   _______________ psi

Final Drill Pipe   = $\dfrac{\text{Kill Mud Weight (\quad) ppg x spp (\quad) psi}}{\text{Original Mud Weight (\quad) ppg}}$
       Pressure

                        =   _______________ psi

Pressure Chart	
Strokes	Pressure

Initial
Drill
Pipe Pressure

Pressure Chart	
Strokes	Pressure

Figure 29-4. Well killing worksheet.

On the prerecorded information section (see Figure 29-4), the reduced circulating pressure in our case will be 30 strokes per minute = 1,000 psi. (See Appendix F for pump output tables.) The surface-to-bit time should be calculated each day and recorded on a chart.

The mud loggers normally keep a record of surface-to-surface time, simply called a "round trip," by dropping a carbide bomb. This is pumped down the hole and when it returns to the surface the number of strokes recorded to "round trip it" will yield washout factors. When a kick occurs, this will enable you to more accurately calculate round trip time at the reduced stroke count. (A good consultant employs a chart as in Figure 29-5.)

An annular capacity of 4.5 XO drill pipe in 10¾ in. casing (check cement manual) equals 3,000 ft surface casing times 0.0784 which equals 235.2 bbl. The capacity of the open hole will be found in the volume and height between drill pipe and open hole sections of the cement book. Since there are 550 ft of drill collars, this figure also needs to be entered. The calculations are as follows:

Since the hole is 9⅞ in., figure the barrels per foot times height, which is 0.0471 bbl/ft times 550 ft, which equals 25.90 bbl. Since the difference is 5,450 ft of open hole that has drill pipe, the barrels equal 0.0751 bbl/ft times 5,450 ft, which comes to 409.29 bbl. Adding the three:

409.29 bbl + 25.90 bbl + 235.2 bbl = 607.39 bbl

Each day this chart needs to be revised in case an emergency arises. This is simple: just multiply the new depth by the figure obtained from the cement book, and add it daily to the chart. On the drill pipe capacity, look in the capacity section of the cement book and obtain the following information:

capacity DP = 0.01422 bbl/ft times 8,450 ft
            = 120.15 bbl (120 bbl rounded off)

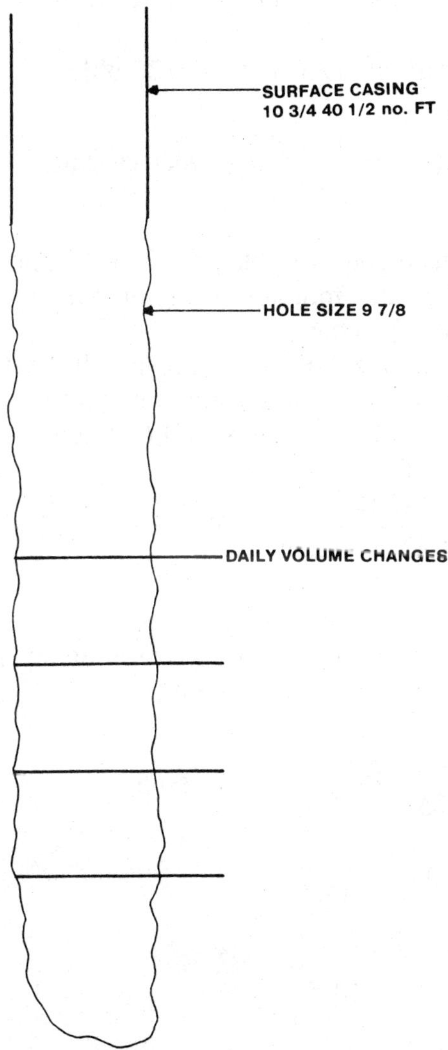

Figure 29-5. The daily volume chart is used to keep track of how much mud volume is in the hole.

$$\text{capacity DC} = 0.0108 \text{ bbl/ft times } 550 \text{ ft}$$
$$= 5.94 \text{ bbl (6 barrels rounded off)}$$

$$\text{total capacity} = 120 + 6 = 126 \text{ bbl}$$

$$\text{pump strokes} = \frac{126 \text{ bbl}}{0.185} = 681 \text{ strokes}$$

Since the reduced pump strokes are 30 per minute, we divide 681 by 30 for 22.7 minutes from surface to bit. This also is recorded on the chart.

Next record the SIDPP—in this case 300 psi. The casing pressure is 450 psi. The pit volume increase is 20 bbl. Next calculate the kill mud increase. The simple formula is:

$$W2 = \frac{\text{SIDPP}}{0.052 \times \text{depth}} + W1$$

where $W1$ = old mud weight
$\quad$ SIDPP = shut-in drill pipe pressure (psi)
$\quad\quad W2$ = new mud weight

$$W2 = \frac{300}{0.052 \times 9{,}000 \text{ ft}} + 10.5$$

$$W2 = \frac{300}{468} + 10.5$$

$$W2 = 0.64 + 10.5$$

$$W2 = 11.14 \text{ (round off to 11.2 ppg)}$$

The new mud weight required is 11.2 ppg.

When you use the wait-and-weight method, plot a graph of the circulating pressures over a period of time. The *initial circulating pressure* (ICP) will be the shut-in drill pipe pressure plus the kill rate pump pressure:

$$SIDPP + KRPP = ICP$$

When the new mud is added, it is heavy enough to replace the shut-in pressure.

The *final circulating pressure* (FCP) is the kill rate pressure taken by the driller each tour. It is corrected for heavier mud. Heavy mud takes more pressure to pump than lighter mud. The graph will start at the beginning of the well kill rate and end when the circulation has started around the bit. Treat all kicks as a gas kick, since gas expands and liquid does not. Most operators use a chart of circulating pressures versus time or pump strokes.

The FCP is the kill rate pump pressure (KRPP) times the kill mud weight (KMW), divided by the present mud weight (PMW):

$$KRPP \times KMW \div PMW = FCP$$

All this should be learned at a blowout school. All I have done is cover the highlights. A blowout school's textbook is as large as this whole book.

In the field most consultants use the kill formula and start kill operations without using the worksheet. Experience teaches the consultant what is needed and how long it takes to kill a well based on depth. But until you can kill a well through experience, use the worksheet if one is available. A consultant must carry a worksheet with him since no worksheets are kept at the rig. Always keep in mind, however, that nothing is accurate in the field (as some engineers would like for you to believe). Once you begin open hole drilling, it's a whole new ball game. Even the pressure gauges on the rig are wrong. After a rig has been torn down and put up and dragged through the mud quite a few times, no gauge will be accurate. Keeping all this in mind when killing a well will make you a better consultant.

Circulate-and-Weight Method

This method is basically the same as the wait-and-weight method except you start mixing and pumping at the same time, instead of building mud weight, then pumping it down.

This method is very good in highly porous zones, since too much weight can cause lost circulation, which leads to more problems. By bringing up the weight two or three points at a time, you can feel your way to the right mud weight. Many times you will find the "kick" is controlled without reaching the kill weight. This is important because the less mud weight in the hole, the better the drilling rate will be and also the less pressure on the surface casing shoe.

Anytime there is a kick, the casing shoe is in danger. For example: if the casing shoe is at 3,000 ft and you tested to a 13.5 EMW, then the maximum psi you can put on the shoe is determined by the following:

$$13.5 \times 0.052 \times 3,000 \text{ ft} = 2,106 \text{ psi}$$

So if the mud weight is at 13 ppg, the maximum pressure you can hold on the casing while handling the kick is:

13.5 EMW for the well is 2,106 psi
13 ppg $\times$ 0.52 $\times$ 3,000 ft
   = 2,028 psi (pressure on the shoe)

To calculate the difference simply subtract the following:

 2,106 psi (maximum tested pressure)
(2,028) psi (pressure on shoe now)
    78 psi (difference)

As can be seen, there is trouble, so keep the superchoke system open all the way. This situation demands full attention from all hands, as the casing seat is about to be lost. (Whenever this kind of trouble occurs, do not get shaken up or at least do not show it in front of the men, as they are depending on you.) Get on the superchoke yourself and make sure that the casing pressure is bled off. If it goes up, open the choke until it goes back to 10 psi. After the kick is circulated out, the pressure will fall to zero when the pumps are turned off.

This is not a normal problem, but it happened to me in southern Louisiana. Every well around ours had blown out, and we took a kick from a transition zone without intermediate casing in the hole. On the final circulation had I not set the choke wide-open, we would have lost the surface shoe. When we logged it, this well was not commercial, so we plugged and abandoned it. And when we pumped the reserve pit down the surface casing, we broke the shoe with a pressure equal to a 13.2 ppg mud weight. Sometimes even with good engineering and blowout schools, it is just luck that saves the day. Had this well lost the shoe, we surely would have had problems. Fast thinking and good Louisiana hands pulled this well through—and one Texan.

If you have maxed out the mud weight and still are having trouble controlling pressure, one way to heal the high pressure zone is to pull one stand of pipe and rehook up the kelly, circulate, and slowly rotate the pipe. This allows the formation to bridge over below the bit and stop the kicking of the formation. This sometimes works, but after you pull one stand of pipe, you need to circulate until the well settles down. This may take 24 hours. (See Figure 29-6.)

Another thing you can do is send a plug of barite or cement downhole, pull two or three stands, reconnect the kelly, and circulate for 24 hours. This sometimes heals the zone. If that does not heal the zone, an intermediate string must be run if drilling is going to continue.

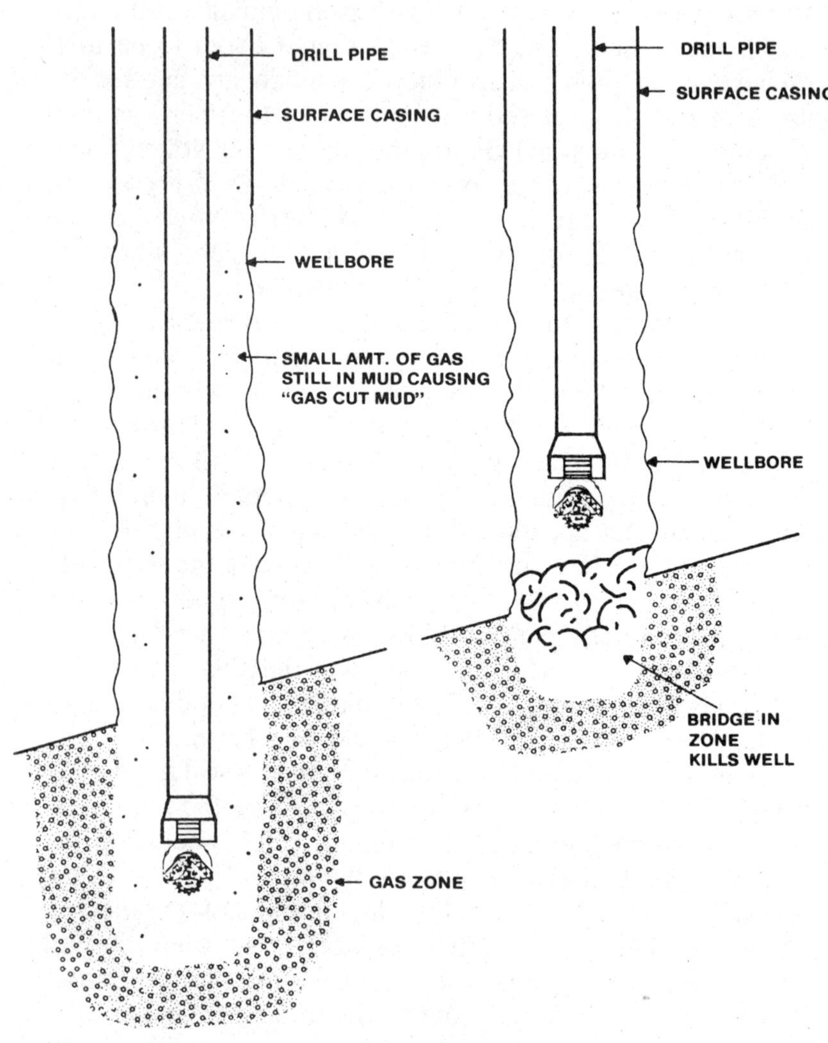

Figure 29-6. Pulling one stand of pipe allows formation to bridge and stops the kick.

On proven fields, kicks are not really a problem because a correlation between the nearest well tells where the pressure will come up, so it's easy to keep the mud weight right when a high-pressure zone is encountered, since the mud can be weighted up before drilling through that zone. On wildcats, keep an eye peeled all the time. Trouble is usually around the corner.

Controlling pressure in a delicate situation requires the teamwork of many men and a consultant who knows his business.

30
Oilfield Firefighting

The most colorful part of the oil business is the fighting of oilwell fires (Figure 30-1). This business was started many years ago by men who risked their lives for money and adventure. Their lifespans were not long, and they lived for the moment. Then the great Myron McKinley formed his company. He traveled all over the world, putting out fires. Some of the old film clips show him driving up to burning wells, wearing a pair of khakis and an Indiana Jones type of hat. He was quite colorful. His ability to find the parts he needed whenever he needed them was amazing. There are stories about his driving up to a well and telling the operator he would be right back, then driving to location with a tanker full of mercury to pump down the well to kill it. Myron became quite wealthy putting out fires.

The next man to make it big is the world-famous Paul ''Red'' Adair. He added colorful red uniforms and a showmanship second to none. His flamboyant style left presidents of big oil companies with their mouths open. He could

Figure 30-1. The scene after a fire has been put out—not much left. (Courtesy of Wild Bunch Hellfighters, San Antonio, TX)

convince the oil companies that fighting fires was not cheap, and if you called him, it would cost you. He has made his mark on the world. Several of his team quit and formed their own companies. History will surely record him as the best firefighter and tactics man in the business. Red Adair's methods were developed by Myron McKinley and him. They go back 40 to 50 years. The methods involved explosives, water, and a lot of equipment. It was quite a show to see.

Basically, the conventional method to kill and cap a well involves first clearing the location and establishing water to be used in the firefighting effort. The location may need a reserve pit and ditches dug to control the oil on location. After all that is done, the firefighters get close to the well and determine the best way to cap it. The necessary parts are ordered from Houston or other places. Next either the firefighters go in, blow the fire, and cap the well, or they let the well bridge off a little so that the pressure and volume are lowered and thus it is simpler to cap it. A lot of the wells bridge over before all the equipment is on location. (See Figure 30-2.) In summary, the conventional methods require water and explosives, and all the firefighting companies use the same methods, created years ago.

In 1976 a company was formed to handle fires and blowouts via good engineering, called the Wild Bunch Blowout and Well Control Company Inc. The founders invented nitrogen firefighting before most people even knew what nitrogen could do. For the first time in history they put out a fire in Wyoming with nitrogen. They invented the nitrogen bomb nozzle, the nitrogen ring (which allows a fire to be raised so workers can work under the fire), the inverted blowout interceptor tool (used on open-hole blowouts), and the nitrogen shield (which allows a worker to work on a

Figure 30-2. North slope action—flaring a well that is under control. (Courtesy of Wild Bunch Hellfighters, San Antonio, TX)

wellhead without being burned). The technology was so new that it changed the industry. In this age of technology and engineering, the Wild Bunch has a team of the sharpest blowout men in the business. They are pushing hard to be the pacesetters of the 1990s. It just makes sense that engineers should call engineers when a well problem arises. The Wild Bunch has the best well-rounded petroleum engineers in the world. Several industry experts call them the ''special forces'' of engineers.

Figure 30-3. A snubbing unit killing a well under pressure. (Courtesy of Wild Bunch Hellfighters, San Antonio, TX)

The firefighting business has been one of successors, and when Red Adair retires, I am sure the Wild Bunch Hellfighters will take up the mantle, until they give it up to more advanced methods or younger men. Right now the Wild Bunch has the edge on technology and youth.

Firefighting Procedures

The standard oilwell fire usually occurs when a well is being drilled. Normally a kick gets out of control for a number of reasons, basically human error or inexperience. If a fire happens on location for whatever reason, it is the consultant's job to vacate the location as quickly and safely as possible. A prearranged plan must be made before a fire occurs.

I have always told my hands on location, "Plan well for a disaster before it happens, and it probably won't happen." I always train my men to look for and be ready for problems and to plan everything around having a disaster, and you will probably never have one. I have never had a problem that resulted in a well being lost in all my years of consulting. Good disaster planning was the reason. It is not macho to take chances with someone's life. Kick drills must be run every so often, and if you hold a surprise drill, then you know how the men will act under fire. But if a well blows out or catches on fire, make sure all the men are accounted for. Call for emergency personnel to take care of any burns, wounds, etc. If communication on location is still intact, call the oil company's engineer and tell him what is going on. Make sure that the toolpusher has located his men. Do not try to be a hero, as you are not paid to fight fires. Too many heroes have died while trying to save a rig. Once the kick is out of control, the consultant's main job is to get the men out and secure the area until emergency help arrives. The engineer will usually call Wild Bunch Hellfighters or some other firefighting company that extinguishes wild wells. When the firefighters arrive on location and sign a contract with the operator, then the dangerous work begins.

First the firefighting company must decide the best approach to use. Equipment must be ordered, and water has to be located in large amounts to be used to cool the area. Often water can be piped onto the location from a tank, pond, river, or irrigation wells, depending on where the largest source is. Some bulldozers may be used to build holding tanks for water, and some tanks may be built to hold loose oil that has not burned. The wellhead area must be cooled off with either the nitrogen bomb nozzle or water. The nitrogen bomb nozzle is faster and cheaper, since no water is needed to cool the area. After the wellhead area is cooled, the firefighters can take a close look at the wellhead and decide what needs to be done. At this point the area has been cleared by bulldozers and cranes. Clearing the rig and all the debris can be dangerous because men and equipment must approach the wellhead. Water is normally sprayed on both the men and the equipment while this cleanup is going on, to keep them cool. If the well is H_2S, extra precautions must be taken. An H_2S service company must be hired to help the firefighters and workers. This company will keep the men from breathing dangerous gases. The saying ''No two wells are the same'' really applies to firefighting. The most unusual things occur.

After the approach has been chosen, the wellhead capping equipment is ordered, while the firefighters prepare the well for capping. This is usually very dangerous, and it takes men who have guts *and* a good understanding of what they are up against. The wellhead can pose all kinds of problems, and the firefighters must decide where to cap and all procedures to be followed before capping. At this point the well can be cooled off with water, to prevent flashback if the well is to be extinguished by explosives. Nitrogen offers the best insurance against flashback, but it is used only by Wild Bunch Hellfighters. Most conventional firefighters use lots of water. The use of nitrogen is new and slow to be accepted by older firefighting firms, but eventually it will become the standard, until someone comes up with a still more advanced system. Next the capping device is installed by the firefighters. A well can be capped while it is burning or while it is flowing, depending on the situation at hand and whether poisonous gases are present. Capping procedures can change from well to well (see Figure 30-3). That is why firefighting costs so much to the operator—it is due to the equipment expense.

After the well is capped, the well can be closed and killed. Then the operator takes over and the firefighters go home.

Some oilwell fires must be dealt with differently because no one can get to the wellhead because of cratering, so a relief well must be drilled. It is expensive to kill a well. Basically another well is drilled, and directional tools are used to get back into the zone where the well kicked. The zone can be pumped with heavy mud and sometimes cement. The hydrostatic effect is used to keep down the pressure in the zone, and that in turn kills the other well. This can ruin the attempt to go back into that zone and make a well. Normally one firefighter supervises the relief well, so nothing else will go wrong, and gives kill orders once the depth is reached.

Leaking valves and flanges can be stopped by pumping different things into the well—pieces of rubber, golf balls, and any flexible material. After the well flow is stopped, several different methods can be used to put it back in production or to continue drilling. For example, if a valve is leaking, golf balls are pumped down the hole; then on their return they will flatten out in the hole, and the pressure will hold and seal them. This may sound very nontechnical, but golf balls have saved many a well from being lost. Pieces of rubber tire have done the same.

Blowouts are the responsibility of the firefighter, too. A blowout is more dangerous than a fire in some respects, because the well could ignite at any time. If a well blows, then the consultant should try to shut it in. If it is impossible to shut it in and be safe, then firefighters should be called in. Remember, they are paid to take chances, and they know the risks; you do not. Under current Workers' Compensation laws, heroes who get hurt receive the same benefits as someone in a regular rig accident. In short, leave the firefighting to the professionals. A blowout is referred to by most firefighters as a well that has not yet caught fire. If you can shut in a well, then start the kill procedures you learned at blowout school. Remember the lives of your crew are worth more than all the rigs running, so make sure in a blowout that you vacate the location. If your well is H_2S and it blows, the decision must be made to shoot the well with a flare gun and set it on fire to burn the poisonous gas. This has to be done if the gas will kill people nearby or your crew. I know of only a few times when that decision had to be made, but losing a rig is better than losing life. The H_2S is simply too dangerous to mess with; let the firefighters deal with the fire. Deciding to burn down a rig is hard, and I am sure the consultant figures it is his last job, but time is important.

Undergound blowouts are another problem that requires firefighters. An underground blowout requires special attention because the loss of circulation can cause the well to get away from the operator. The firefighters' function is to assess the extent of the problem and advise the operator on the approach to control the well and seal the formation. Usually a mud engineer is called to the location to assist the firefighers. Usually the formation is sealed, and then the right mud weight is used to stop the well from kicking, all the while attempting not to lose circulation. In some cases a relief well has been drilled to assist the rig in trouble. This is rare, however. Once an underground blowout is brought under control, usually the operator tries to run pipe and finish the well.

One problem that has plagued the firefighting industry is the reluctance of operators to try new technology in fighting fires. The conventional methods are primitive and do not reflect much engineering thought.This problem has really come into the light in the recent Kuwait situation. First, there is a shortage of firefighting companies in the world, and this is the result of operators not giving new companies a chance to prove themselves. This has created a shortage of firefighters and blowout specialists. Most of the older oil companies are still run by engineers trained in outdated technology. If a new idea is presented to them, they are afraid to try something unknown. I have always prided myself on trying new ideas and new companies, to see if they can perform, and most of the time it has been a good decision. I used the first security shock sub in Texas, among several other new items, and the result was good. If we do not embrace new techniques and technology and if other, new companies are not tried, then we will have the same situation as in Kuwait in 1991—no hands and slow work! If an engineer and some hands start a firefighting company, and they have good oilfield backgrounds, good ideas, and good technology, why not give them a try? The industry has been controlled by nonengineers and roughnecks for too long.

In summary, have a good blowout and fire plan on the rig. Prepare for disaster and it probably will not happen.

31
Completing a Horizontal Well

Earlier we discussed how to set a packer in the Austin chalk horizontal well. However I thought a separate chapter on completions and some of my predictions in the future should be discussed even though this is not a book on well completions but drilling.

A lot of wells will be drilled in other areas, not just in the chalk, and completions will not be open-hole as in the chalk, but will require the use of liners and in some cases production strings. The biggest problem is the cementing of the horizontal portions. The cuttings in the horizontal portion of the well will tend to lie on the bottom portion of the wellbore, creating a problem when the cement is pumped. Most wells at this point have not been successful in separating the zones of oil and water, and after a short time more water is produced than oil or gas. Hopefully some mud company or cement company will invent a wash to remove the loose cuttings before the horizontal zone is cemented. Polomar and standard muds have not worked. If asbestos were not illegal, a good Visbestos sweep might do the trick. I have found that when something works well, it is usually against OSHA or state agency regulations. Maybe a liquid mud, already mixed, containing some asbestos or similar material could be brought to the location and then pumped down the hole.

Another problem is that the centralizers in current use are not strong enough to keep the pipe off the bottom. Special centralizers will have to be built so that the pipe can slide into position and be off the bottom, for the cement to do its job. If these special centralizers are run into the horizontal portion slowly, a big problem is solved.

Several companies have designed a tool to turn the pipe after it is in place and cement has been pumped, but its success rate has been poor. I know of several occasions when the tool broke during the operation. The centralizers will also need to be able to slip while the pipe is being rotated. Since this is new technology, horizontal wells will be the new horizon for oilmen. Given the ability to pull off a good cement job, horizontal wells could take off like barn burners. Of course, in the chalk completions have been successful; but even in the chalk it would be nice to isolate and produce out of fractures until they are depleted and then move on to the next fracture. Control of nature is the main concern of oilmen, and with open-hole wells that have not been logged, we work on blind faith.

Logging techniques will also have to be improved. Right now it is expensive to log a horizontal well, and failure is common. I am sure logging companies are working around the clock to improve and give us a better system.

One of the things I have begged operators to try is the use of multiple horizontal wells out of one well. Think how profitable one well would be with four 3,000-ft horizontal drilled directions. Instead of drawing from one direction, you could cover a circular area 3,000 ft from the wellbore.

To me that makes sense, and I am sure that it will be the wave of the future. A device could be built to go downhole to the bottom of the vertical well to allow four different directions and running of four liners or four open holes. The only thing that needs to be negotiated is the production level permitted by the state. With four horizontal wells, production would be great. Money and fear of failure have stopped most operators from venturing into the unknown. The excitement created in the chalk by your horizontal well could be multiplied four times, and with the future rotating heads and cheaper technology, the well could virtually pay for itself before it was finished. A well could flow during the total operation, and the driller could strip in and out of the pressure and keep up a good head of steam. Once MWD tools and better motors are developed that will go hundreds of hours, two to four holes could be drilled without tripping. This will be the future of horizontal wells in the chalk. In other zones tripping will be necessary because liners have to be set before the next horizontal direction is drilled. In the chalk a small cement plug could be set to reopen after the other three are drilled. Then there is a fast trip into the well to redrill out the plugs and set the packer and tubing. The well could be kicked off by the completion man. Accurate measurements, of course, will be the secret to the success of the well.

The future of horizontal drilling will include the coiled tubing units. The tubing will be attached to a motor. These units are the future, and you will see them being used soon. And when they are, look out conventional drilling!

Many existing wells could be reentered, a window cut out at different depths, and liners set while moving up the hole in different directions. The possibilities are endless, and I hope oilmen will be adventuresome and give my future predictions a good shot. For myself, before horizontal wells came about, drilling vertical wells had become quite boring, except when I was in high-pressure zones. With horizontal drilling the thrill is back for the oilman drilling oil or gas wells. Fighting kicks and controlling wild wells are what I enjoy in the oil business, and I hope that after horizontal wells have been tamed by technology, some other exciting way to drill for oil will be discovered. Maybe wells will be drilled on other planets in the not-so-distant future. I have my hard hat and spacesuit packed.

Appendix A
*IADC Footage Drilling Contract**

*Courtesy of the International Association of Drilling Contractors

INTERNATIONAL ASSOCIATION OF DRILLING CONTRACTORS
DRILLING BID PROPOSAL
AND
FOOTAGE DRILLING CONTRACT — U.S.

TO: ____

____

____

____

____

Please submit bid on this drilling contract form for performing the work outlined below, upon the terms and for the consideration set forth, with the understanding that if the bid is accepted by

____

this instrument will constitute a contract between us. Your bid should be mailed or delivered not later than _________________________ P.M.

on _________________________, 19 _________ to the following address:

____

____

★   ★   ★   ★   ★   ★

THIS AGREEMENT, made and entered into on the date hereinafter set forth by and between the parties herein designated as "Operator" and "Contractor".

OPERATOR: ____

Address: ____

____

CONTRACTOR: ____

Address: ____

(continued on next page)

U.S. Footage Contract continued

IN CONSIDERATION of the mutual promises, conditions and agreements herein contained and the specifications and special provisions set forth in Exhibit "A" and Exhibit "B" attached hereto and made a part hereof, Operator engages Contractor as an Independent Contractor to drill the hereinafter designated well in search of oil or gas on a footage basis.

For purposes hereof the term "footage basis" means Contractor shall furnish the equipment, labor, and perform services as herein provided to drill a well, as specified by Operator, to the contract footage depth. Subject to terms and conditions hereof, payment to Contractor at a stipulated price per foot of hole drilled is earned upon attaining such contract footage depth or other specified objective. While drilling on a footage basis Contractor shall direct, supervise and control drilling operations and assumes certain liabilities to the extent specifically provided for herein. Notwithstanding that this is a footage basis contract, Contractor and Operator recognize that certain portions of the operations as hereinafter designated, both above and below contract footage depth, will be performed on a daywork basis. For purposes hereof the term "daywork basis" means Contractor shall furnish equipment, labor, and perform services as herein provided, for a specified sum per day under the direction, supervision and control of Operator (which term is deemed to include any employee, agent, consultant, or subcontractor engaged by Operator to direct drilling operations). When operating on a daywork basis, Contractor shall be fully paid at the applicable rates of payment and assumes only the obligations and liabilities stated herein as being applicable during daywork operations. Except for such obligations and liabilities specifically assumed by Contractor, Operator shall be solely responsible and assumes liability for all consequences of operations by both parties while on a daywork basis, including results and all other risks or liabilities incurred in or incident to such operations.

1. LOCATION OF WELL:

Well Name
and Number: ____

Parish/
County: _________________________ State: ___________________ Field
Name: ____________________________________

Well location and
land description: ____

____

The above is for well and contract identification only and Contractor assumes no liability whatsoever for a proper survey or location stake on Operator's lease.

2. COMMENCEMENT DATE:

Contractor agrees to use best efforts to commence operations for the drilling of well by the ___________________________________ day of ___________________________, 19 _________, or ____.

(continued on next page)

3. DEPTH:

Subject to the right of Operator to direct the stoppage of work at any time (as provided in Par. 6), the well shall be drilled to the depth as specified below:

3.1 Contract Footage Depth: The well shall be drilled to _______________ feet or ____ formation, or to the depth at which the ___________ inch casing (oil string) is set, whichever depth is first reached, on a footage basis and Contractor is to be paid for such drilling at the footage rate specified below, which depth is hereinafter referred to as the contract footage depth.

3.2 Daywork Basis Drilling: All drilling below the above specified contract footage depth shall be on a daywork basis as defined herein and Contractor shall be paid for such drilling at the applicable daywork rate specified below.

3.3 Complete Daywork Basis Drilling: If all operations hereunder are performed at applicable daywork rates, provisions of this contract applicable to drilling on a "footage basis" shall not apply.

3.4 Maximum Depth: Contractor shall not be required to drill said well under the terms of this contract below a maximum depth of _______________ feet.

4. FOOTAGE RATE, DAYWORK RATES, BASIS OF DETERMINING AMOUNTS PAYABLE TO CONTRACTOR:

Contractor shall be paid at the following rates for the work performed hereunder.

4.1 Footage Rate: For work performed on a footage basis the rate will be $ ___________________ per linear foot of hole drilled determined by steel line measurement from the surface of the ground if Contractor provides cellar, or from the bottom of the cellar if Operator provides cellar, less footage made in regular size hole while working on daywork basis.

4.2 Operating Day Rate: For work performed on a daywork basis the daywork rate per twenty-four hour day with _____________________________ man crew shall be:

Depth Intervals

From	To	Without Drill Pipe	With Drill Pipe
_______________	_______________	$_____________________ per day	$_____________________ per day
_______________	_______________	$_____________________ per day	$_____________________ per day
_______________	_______________	$_____________________ per day	$_____________________ per day

Using Operator's drill pipe $___________________ per day.

(continued on next page)

U.S. Footage Contract continued

If under the above column "With Drill Pipe" no day rates are specified, the daywork rate per twenty-four hour day when drill pipe is in use shall be the applicable daywork rate specified in the column "Without Drill Pipe" plus compensation for any drill pipe actually used at the rates specified below, computed on the basis of the maximum drill pipe in use at any time during each twenty-four day.

DRILL PIPE RATES PER 24-HOUR DAY

Straight Hole	Size	Grade	Directional or Uncontrollable Deviated Hole	Size	Grade
$______________ per ft.	__________	__________	$______________ per ft.	__________	__________
$______________ per ft.	__________	__________	$______________ per ft.	__________	__________
$______________ per ft.	__________	__________	$______________ per ft.	__________	__________

Drill pipe shall be considered in use not only when in actual use but also while it is being picked up or laid down. When drill pipe is standing in the derrick, it shall not be considered in use, provided, however, that if Contractor furnishes special strings of drill pipe, drill collar, and handling tools as provided for in Exhibit "A", the same shall be considered in use at all times when on location or until released by Operator. In no event shall fractions of an hour be considered in computing the amount of time drill pipe is in use but such time shall be computed to the nearest hour, with thirty minutes or more being considered a full hour and less than thirty minutes not to be counted.

4.3 Work Stoppage Rate: $______________ Per Day $______________ Per Hour

The above rate shall apply under the following circumstances:

(a) During any continuous period that normal operations are suspended or cannot be carried on due to conditions of force majeure as defined in Paragraph 21 hereof. It is understood, however, that Operator shall have the right to release the rig in accordance with Operator's right to direct stoppage of the work (See Paragraph 6), effective when conditions will permit the rig to be moved from the location.

(b) During any period when Contractor has notified Operator that the rig is available for movement to the drilling site and movement cannot be accomplished because of Operator's failure or inability to furnish and/or maintain adequate roadway and/or canal to location and/or location and/or weather prevents positioning the rig on a water location drill site.

(c) During any period after operations under this Contract have been completed and Operator has released the rig and the same cannot be dismantled and/or transported from the location due to inadequate roadway or canal or weather or water conditions which will not allow such activity to be conducted with reasonable safety.

(d) Operator agrees at all times to maintain the road and location in such a condition that will allow free access and movement to and from the drilling site in an ordinarily equipped highway type vehicle. If Contractor is required to use dozers, tractors, four-wheel drive vehicles, or any other specialized transportation equipment for the movement of necessary personnel, machinery, or equipment over access roads or on the drilling location, Operator shall furnish the same at his expense and without cost to Contractor. The actual cost of repairs to any transportation equipment furnished by Contractor or his personnel damaged as result of improperly maintained access roads or location will be charged to Operator.

(continued on next page)

U.S. Footage Contract continued

4.4 Repair Time: In the event it is necessary to shut down Contractor's rig for repairs excluding routine rig servicing while Contractor is performing daywork hereunder, Contractor shall be allowed compensation at the applicable daywork rate for such shut down time up to a maximum of ______________ hours for any one repair job.

4.5 Standby Time Rate with Crews: $______________ per twenty-four (24) hour day. Standby time shall be defined to include time when the rig is shut down although in readiness to begin or resume operations but Contractor is waiting on orders of Operator or on materials, services or other items to be furnished by Operator.

4.6 Reimbursable Costs: Operator shall reimburse Contractor for the costs of material, equipment, work or services which are to be furnished by Operator as provided for herein but which for convenience are actually furnished by Contractor at Operator's request, plus ______________ percent for such cost of handling.

4.7 Daywork Operations: In addition to other work specified herein the following work performed by Contractor shall be on a daywork basis:

(a) All drilling below the contract footage depth as provided in Par. 3.1, including the setting of any string of casing below such depth;

(b) All work performed by Contractor, whether or not prior to reaching the contract footage depth, in an effort to restore the hole to such condition that further drilling or other operations may be conducted, in the event of loss or damage to the hole as a result of the failure of Operator's casing or equipment either during or after the running and setting of such casing or as a result of the subsequent failure of the cementing job resulting in parted casing;

(c) All work performed when conditions set forth in Paragraph 12 are applicable;

(d) All other work performed by Contractor at the request of Operator, regardless of depth, which is not within the scope of the work to be performed on a footage basis, including all coring, drill stem testing, bailing, gun or jet perforating, electric logging, acid treatment, shooting, cleaning out, hydraulic fracturing, plugging, running tubing, setting liners, squeeze cementing, abandoning well and installation of well head equipment.

4.8 Daywork Time: In determining the amount of daywork time for which Contractor is to be compensated at the applicable daywork rate, it is agreed that such daywork time shall begin when Contractor in accordance with the terms hereof, suspends normal footage drilling operations. There shall be included in daywork time any time required to condition the hole preparatory to performing such daywork and also the time required to restore the hole to the same drilling conditions which existed when operations were suspended for the purpose of beginning daywork, in order to again resume normal footage drilling operations.

4.9 Revision in Rates: The rates and/or payments herein set forth due to Contractor from Operator shall be revised to reflect the change in costs if the costs of any of the items hereinafter listed shall vary by more than ______________ percent from the costs thereof on the date of this Contract or by the same percent after the date of any revision pursuant to this paragraph:

(a) Labor costs, including all benefits, of Contractor's personnel;
(b) Contractor's cost of insurance premiums;
(c) Contractor's cost of fuel, the cost per gallon/MCF being $______________;

(continued on next page)

U.S. Footage Contract continued

(d) Contractor's cost of catering, when applicable;

(e) If Operator requires Contractor to increase or decrease the number of Contractor's personnel;

(f) Contractor's cost of spare parts and supplies with the understanding that such spare parts and supplies constitute __________ percent of the operating Rate and that the parties shall use the U.S. Bureau of Labor Statistics Oilfield Machinery and Equipment Wholesale Price Index (Code No. 1191-02) to determine to what extent a price variance has occurred in said spare parts and supplies;

(g) If there is any change in legislation or regulations in the area in which Contractor is working that alters Contractor's financial burden.

5. TIME OF PAYMENT:

Subject to Operator's right to require Contractor to furnish him with satisfactory evidence that Contractor has paid all labor and material claims chargeable to Contractor, payment becomes due by Operator to Contractor as follows:

5.1 Footage Basis: If the well is drilled to total depth on a footage basis, payment becomes due for all services (footage and daywork) when Contractor completes the performance of the services which he agrees to perform under this contract, provided, however, if Contractor prior to the completion of the contract performs a substantial amount of daywork, payment for such daywork shall be due and payable upon presentation of invoice therefor at the end of the month in which such daywork was performed.

5.2 Daywork Basis: If the entire hole or the bottom section of the hole is drilled on a daywork basis, payment shall become due as follows: Upon Contractor's completion of the footage basis drilling to the depth specified above and upon acceptance by the Operator of the hole as drilled to such depth in accordance with this agreement, payment becomes due for all footage drilled and for all work performed on a daywork basis to the date of completion of the footage drilled. Payment for drilling and other work performed at daywork rates below the depth specified at which daywork basis drilling commences shall become due upon acceptance by Operator of the work performed in accordance with this contract upon presentation of invoice therefor upon completion of the well or at the end of the month in which such daywork was performed, whichever shall first occur.

5.3 Disputed Invoices and Late Payment: Operator shall pay all invoices within __________ days after receipt except that if Operator disputes an invoice or any part thereof, Operator shall, within fifteen days after receipt of the invoice, notify Contractor of the item disputed, specifying the reason therefor, and payment of the disputed item may be withheld until settlement of the dispute, but timely payment shall be made of any undisputed portion. Any sums (including amounts ultimately paid with respect to a disputed invoice) not paid within the above specified days shall bear interest at the rate of __________ percent or the maximum legal rate, whichever is less, per month from the due date until paid. If Operator does not pay undisputed items within the above stated time, Contractor may terminate this Contract as specified under subparagraph 6.2.

5.4 Attorney's Fees: If this Contract is placed in the hands of an attorney for collection of any sums due hereunder, or suit is brought on same, or sums due hereunder are collected through bankruptcy or probate proceedings, then Operator agrees that there shall be added to the amount due reasonable attorney's fees and costs.

(continued on next page)

U.S. Footage Contract continued

6. STOPPAGE OF WORK BY OPERATOR OR CONTRACTOR:

6.1 By Operator: Notwithstanding the provisions of Paragraph 3 with respect to the depth to be drilled, Operator shall have the right to direct the stoppage of the work to be performed by the Contractor hereunder at any time prior to reaching the specified depth, and even though Contractor has made no default hereunder, and in such event Operator shall be under no obligation to Contractor except as set forth in subparagraph 6.3 hereof.

6.2 By Contractor: Notwithstanding the provision of Paragraph 3 with respect to the depth to be drilled, in the event of Force Majeure necessitating a termination of operations, or in the event of total or constructive total loss of the rig, or if Operator shall become insolvent, or be adjudicated a bankrupt, or

file, by way of petition or answer, a debtor's petition or other pleading seeking adjustment of Operator's debts, under any bankruptcy or debtor's relief laws now or hereafter prevailing, or if any such be filed against Operator, or in case a receiver be appointed of the Operator or Operator's property, or any part thereof, or Operator's affairs be placed in the hands of a Creditor's Committee, or, following ten days written notice to Operator, if Operator does not pay Contractor within the time specified in subparagraph 5.3, all undisputed items due and owing, Contractor may, at his option, elect to terminate further performance of any work under this contract and Contractor's right to compensation shall be as set forth in subparagraph 6.3 hereof. In addition to Contractor's right to terminate performance hereunder, Operator hereby expressly agrees to protect, indemnify and save Contractor harmless from any claims, demands and causes of action, including all costs of defense, in favor of Operator, Operator's joint ventures, or other parties arising out of any drilling commitments or obligations contained in any lease, farmout agreement or other agreement, which may be affected by such termination of performance hereunder.

6.3 Early Termination Compensation:

(a) Prior to Commencement: In the event this Contract is terminated prior to commencement of operations hereunder, Operator shall pay Contractor as liquidated damages and not as a penalty a sum equal to the Footage Rate (Article 4.1) multiplied by the Contract Footage Depth (Article 3.1) plus the Standby Rate (Article 4.5) for a period of ____________ days for estimated daywork drilling below Contract Footage Depth; or a lump sum of $____________.

(b) Prior to Spudding: If such work stoppage occurs after commencement of operations but prior to the spudding of the well, Operator shall pay to Contractor the sum of the following: (1) all expenses reasonably and necessarily incurred and to be incurred by Contractor by reason of the contract and by reason of the premature stoppage of the work, excluding, however, expenses of normal drilling crew and supervision; (2) ten percent (10%) of the amount of such reimbursable expenses; and (3) a sum calculated at the standby rate for all time from the date upon which Contractor commences any operations hereunder down to such date subsequent to the date of work stoppage as will afford Contractor reasonable time to dismantle his rig and equipment.

(c) Subsequent to Spudding: If such work stoppage occurs after the spudding of the well, Operator shall pay the Contractor (1) the amount owing Contractor at the time of such work stoppage under the footage rate, applicable daywork rate, and standby rate; but in such event Operator shall pay Contractor for a minimum footage of ____________ feet regardless of whether or not the well has been drilled to such depth at the time of work stoppage; or (2) at the election of Contractor and in lieu of the foregoing Operator shall pay Contractor for all expenses reasonably and necessarily incurred and to be incurred by Contractor by reason of this contract and by reason of the premature stoppage of work plus the sum of $____________.

(continued on next page)

U.S. Footage Contract continued

7. CASING PROGRAM:

7.1 The casing program to be followed in the drilling of said well is set forth in Exhibit "A", and Contractor shall drill a hole of size specified in Exhibit "A" to set at the approximate depth therein indicated the size of casing so specified. The exact setting depths for each string of casing shall be specified by Operator. Operator may modify said casing program provided any modification thereof which materially increases Contractor's hazards or costs of performing his obligations hereunder can only be made by mutual consent of Contractor and Operator.

7.2 The setting of any string of casing within the footage contract depth shall be performed as specified in Exhibit "A".

7.3 The setting of any string of casing below the footage contract depth shall be performed by Contractor under the direction of Operator but Operator shall pay Contractor for all time so consumed at the applicable daywork rate.

7.4 Operator reserves the right to require Contractor to set strings of casing or liners in addition to those listed (subject to the limitations upon Operator's right to modify the casing program as provided for in Par. 7.1) and in such event Contractor agrees to provide rig time for cementing and testing cement on such liners and strings of casing and to provide rig time for performing cement squeezing jobs as required by Operator. Operator shall pay Contractor for time consumed by such work at the applicable daywork rate.

8. LABOR, EQUIPMENT, MATERIALS, SUPPLIES, AND SERVICES:

The furnishing of labor, equipment, appliances, materials, supplies, and services of whatever character necessary or proper in the drilling and completion of said well and not otherwise specifically provided for herein shall be furnished by Contractor or Operator as specified in Exhibit "A".

9. DRILLING METHODS AND PRACTICES:

9.1 Contractor shall maintain well control equipment in good condition at all times and shall use all reasonable means to control and prevent fires and blow-outs and to protect the hole.

9.2 Subject to the terms hereof, at all times during the drilling of the well, Operator shall have the right to control the mud program, and the drilling fluid must be of a type and have characteristics acceptable to Operator and be maintained by Contractor in accordance with the specifications shown in Par. 2 of Exhibit "A". No change or modification of said specifications which materially increases Contractor's hazards or costs of performing his obligations hereunder shall be made by Operator without consultation with and consent of Contractor. Operator shall have the right to make any tests of the drilling fluid which may be necessary. Should no mud control program be specified by Operator in Exhibit "A", Contractor shall have the right to determine the mud program and the type and character of drilling fluid during the time that Contractor is performing work upon a footage basis under the terms of this contract.

9.3 Contractor shall measure the total length of drill pipe in service with a steel tape at the point where the contract footage depth has been reached; and when requested by Operator, before setting casing or liner and after reaching final depth.

(continued on next page)

U.S. Footage Contract continued

9.4 Contractor agrees to furnish equipment, workmen and instruments acceptable to Operator and to make slope tests as provided in Exhibit "A". Unless operations are on a daywork basis, all such slope tests shall be made at Contractor's sole risk, cost and expense. If, in the opinion of Operator, it becomes advisable to obtain the use of an additional slope test instrument and accessory equipment for the purpose either of checking previous readings or of determining the direction of the drift, the rental charges therefor shall be paid by Operator, and the running of same shall be on a day-work basis. Should the hole at any depth during the time Contractor is performing work on a footage basis, have either a deviation from vertical or a change in over-all angle in excess of the limits prescribed in Exhibit "A", Contractor agrees to restore the hole to a condition suitable to Operator either by conventional methods and procedures while drilling ahead or by cementing off and redrilling. While operations are being performed on a "Daywork Basis", or during "Complete Daywork Basis Drilling", Contractor agrees to exercise due diligence and care to maintain the straight hole specifications, if any, set forth in Par. 3 of Exhibit "A" but all risk and expense of maintaining such specifications or restoring the hole to a condition suitable to Operator shall be assumed by Operator.

9.5 Contractor will conduct operations to comply with all laws, rules, orders, and regulations, Federal, State, and Local, which are applicable to Contractor, Contractor's business, equipment, and personnel engaged in operations covered by this Contract, including but not limited to those set forth in Exhibit "B". In the event any provision of this Contract is inconsistent with or contrary to any applicable Federal, State or Municipal law, rule or regulation, said provision shall be deemed to be modified to the extent required to comply with said law, rule or regulation, and as so modified said provision and this Contract shall continue in full force and effect.

10. COMPLETION TESTS AND INSTALLATION OF WELL CONNECTIONS OR ABANDONMENT:

Contractor will either complete the well and install well head equipment and connections or plug and abandon same, in accordance with Operator's instructions, at the applicable rates set forth in Par. 4 above, using equipment, materials and services to be furnished and paid for by either Operator or Contractor as specified in Exhibit "A".

11. CORING AND CUTTINGS:

11.1 As directed by Operator and utilizing the type of coring equipment specified and furnished as shown in Exhibit "A", Contractor agrees at any time to take either rat-hole or full hole conventional or wire line cores in the manner requested by Operator. Regardless of depth, all coring shall be paid for at the applicable daywork rate. All coring footage shall be deducted from the total footage charge if the well is being drilled on footage basis at that depth. Reaming of the rat-hole shall be paid for at the applicable daywork rate.

11.2 When requested by Operator, Contractor shall save and identify the cuttings and cores, free from contamination, and place them in separate containers which shall be furnished by Operator; such cuttings and cores shall be made available to a representative of Operator at the location.

12. FORMATIONS DIFFICULT OR HAZARDOUS TO DRILL:

12.1 In the event chert, pyrite, quartzite, granite, igneous rock or other impenetrable substance, is encountered while drilling on the footage basis and the footage drilled during each twenty-four (24) hour period multiplied by the footage rate does not equal the applicable daywork rate plus cost of bits, all drilling operations shall be conducted on a daywork basis at the applicable daywork rate, with Operator furnishing the bits, until normal drilling operations and procedures can be resumed. The footage drilled on daywork rate shall be deducted from the footage charge.

(continued on next page)

U.S. Footage Contract continued

12.2 In the event water flow, domal, steeply dipping or faulted formation, abnormal pressure, underground mine or cavern, heaving formation, salt or other condition is encountered which makes drilling abnormally difficult or hazardous, causes sticking of drill pipe or casing, or other difficulty which precludes drilling ahead under reasonably normal procedures, Contractor shall, in all such cases, without undue delay, exert every reasonable effort to overcome such difficulty. When such condition is encountered, further operations shall be conducted on a daywork basis at the applicable daywork rate until such conditions have been overcome and normal drilling operations can be resumed. Operator shall assume the risk of loss of or damage to the hole and to Contractor's equipment in the hole from the time such condition is encountered. The footage drilled while on daywork basis shall be deducted from the footage charge.

12.3 In the event loss of circulation or partial loss of circulation is encountered, Contractor shall, without undue delay, exert every reasonable effort to overcome such difficulty. When such condition is encountered, Operator shall assume risk of loss of or damage to the hole and to Contractor's equipment in the hole. Should such condition persist in spite of Contractor's efforts to overcome it, then after a period of __________ hours time consumed in such efforts, further operations shall be conducted on a daywork basis at the applicable daywork rate until such condition has been overcome and normal drilling operations can be resumed. The total rig time furnished by Contractor under the terms of this paragraph shall be limited to a cumulative __________ hours. The footage drilled while on daywork basis shall be deducted from the footage charge.

13. REPORTS TO BE FURNISHED BY CONTRACTOR:

13.1 Contractor shall keep and furnish to Operator an accurate record of the work performed and formations drilled on the IADC-API Daily Drilling Report Form or other form acceptable to Operator. A legible copy of said form signed by Contractor's representative shall be furnished by Contractor to Operator.

13.2 Delivery tickets, if requested by Operator, covering any material or supplies furnished by Operator shall be turned in each day with the daily drilling report. The quantity, description, and condition of materials and supplies so furnished shall be checked by Contractor and such tickets shall be properly certified by Contractor.

14. INGRESS AND EGRESS TO LOCATION:

Operator hereby assigns to Contractor Operator's rights of ingress and egress with respect to the tract of land where the well is to be located for the performance by Contractor of all work contemplated by this contract. Should Contractor be denied free access to the location for any reason not reasonably within Contractor's control, any time lost by Contractor as a result of such denial shall be paid for at the applicable rate in keeping with the stage of operations at that time. In the event there are any restrictions, conditions, or limitations in Operator's lease which would affect the free right of ingress and egress to be exercised by Contractor hereunder, its employees, or subcontractors, Operator agrees to timely advise Contractor in writing with respect to such restrictions, conditions, or limitations, and Contractor agrees to observe same.

(continued on next page)

U.S. Footage Contract continued

15. RESPONSIBILITY FOR A SOUND LOCATION:

Operator shall prepare a sound location, adequate in size and capable of properly supporting the drilling rig, and shall be responsible for a conductor pipe program adequate to prevent soil and sub-soil washout. It is recognized that Operator has superior knowledge of the location and access routes to the location, and must advise Contractor of any sub-surface conditions, or obstructions (including, but not limited to, mines, caverns, sink holes, streams, pipelines, power lines and telephone lines) which Contractor might encounter while en route to the location or during operations hereunder. In the event sub-surface conditions cause a cratering or shifting of the location surface, or if seabed conditions prove unsatisfactory to properly support the rig during marine operations hereunder, and loss or damage to the rig or its associated equipment results therefrom, Operator shall, without regard to other provisions of this contract, reimburse Contractor to the extent not covered by Contractor's insurance, for all such loss or damage including payment of work stoppage rate during repair and/or demobilization if applicable.

16. INSURANCE:

During the life of this contract, Contractor shall at Contractor's expense maintain, with an insurance company or companies authorized to do business in the state where the work is to be performed and satisfactory to Operator, insurance coverages of the kind and in the amounts set forth in Exhibit "A". Contractor shall, if requested to do so by Operator, procure from the company or companies writing said insurance a certificate or certificates satisfactory to Operator that said insurance is in full force and effective and that the same shall not be cancelled or materially changed without ten (10) days prior written notice to Operator. For liabilities assumed hereunder by Contractor, its insurance shall be endorsed to provide that the underwriters waive their right of subrogation against Operator. Operator will, as well, cause its insurer to waive subrogation against Contractor for liability it assumes.

17. PAYMENT OF CLAIMS:

Contractor agrees to pay all claims for labor, material, services, and supplies to be furnished by Contractor hereunder, and agrees to allow no lien or charge to be fixed upon the lease, the well, or other property of Operator or the land upon which said well is located.

18. RESPONSIBILITY FOR LOSS OR DAMAGE:

18.1 Contractor's Surface Equipment: Contractor shall assume liability at all times, regardless of whether the work is being performed on a footage basis or daywork basis, for damage to or destruction of Contractor's surface equipment, including but not limited to all drilling tools, machinery, and appliances, for use about the surface, regardless of when or how such damage or destruction occurs, except for such loss or damage as provided in Paragraph 15 and 18.4 herein, and Operator shall be under no liability to reimburse Contractor for any such loss.

18.2 Contractor's In-Hole Equipment — Footage Basis: Contractor shall assume liability at all times while work is being performed on a footage basis for damage to or destruction of Contractor's in-hole equipment, including but not limited to, drill pipe, drill collars, and tool joints, and Operator shall be under no liability to reimburse Contractor for any such loss except as provided for in Paragraphs 12.2, 12.3, 15 and 18.4.

18.3 Contractor's In-Hole Equipment — Daywork Basis: Operator shall assume liability at all times for damage to or destruction of Contractor's in-hole equipment, including but not limited to, drill pipe, drill collars, and tool joints, and Operator shall reimburse Contractor for the value of any such loss or

(continued on next page)

damage; the value to be determined by agreement between Contractor and Operator as current repair cost or ________________ percent of current new replacement cost of such equipment delivered to the well site.

18.4 Contractor's Equipment — Environmental Loss or Damage: Notwithstanding the provisions of Paragraph 18.1 above, Operator shall assume liability at all times for damage to or destruction of Contractor's equipment caused by exposure to highly corrosive or otherwise destructive elements, including those introduced into the drilling fluid.

18.5 Operator's Equipment: Operator shall assume liability at all times for damage to or destruction of Operator's equipment, including but not limited to casing, tubing, well head equipment, and platform if applicable, and Contractor shall be under no liability to reimburse Operator for any such loss or damage.

18.6 The Hole — Footage Basis: Subject to the provisions of Paragraphs 12 and 15 hereof should a fire or blowout occur or should the hole for any cause attributable to Contractor's operations be lost or damaged while Contractor is engaged in the performance of work hereunder on a footage basis, all such loss of or damage to the hole shall be borne by Contractor; and if the hole is not in condition to be carried to the contract depth as herein provided, Contractor shall, if requested by Operator, commence a new hole without delay at Contractor's cost; and the drilling of the new hole shall be conducted under the terms and conditions of this contract in the same manner as though it were the first hole. In such case Contractor shall not be entitled to any payment or compensation for expenditures made or incurred by Contractor on or in connection with the abandoned hole, except for daywork earned in coring, testing, and logging said well for which Contractor would have been compensated had such hole not been junked and abandoned.

Notwithstanding the foregoing provisions, if the hole is lost or damaged as a result of the failure of Operator's casing or equipment either during or after the running and setting of such casing, or as a result of subsequent failure of the cementing job resulting in parted casing, such loss shall be borne by Operator and Contractor shall nevertheless be paid: (a) For all footage drilled and other work performed by Contractor prior thereto; (b) For work performed in an effort to restore the hole to such condition as that further drilling or other operations may be conducted at the applicable daywork rate; and (c) The cost of dismantling the rig and moving to and rigging up Contractor's equipment prior to starting the drilling of a new hole at a location designated by Operator if such be required. The work of drilling the new hole shall be performed by Contractor under the terms and conditions of this contract.

18.7 The Hole — Daywork Basis: In the event the hole should be lost or damaged, while Contractor is working on a daywork basis, Operator shall be solely responsible for such damage to or loss of the hole, including the casing therein, as well as for cost of control of any wild well.

18.8 Underground Damage: Operator agrees to defend and indemnify Contractor for any and all claims against Contractor resulting from operations under this contract on account of injury to, destruction of, or loss or impairment of any property right in or to oil, gas, or other mineral substance or water, if at the time of the act or omission causing such injury, destruction, loss, or impairment, said substance had not been reduced to physical possession above the surface of the earth, and for any loss or damage to any formation, strata, or reservoir beneath the surface of the earth.

(continued on next page)

U.S. Footage Contract continued

18.9 Inspection of Materials Furnished by Operator: (a) Contractor agrees to visually inspect all materials furnished by Operator before using same and to notify Operator of any apparent defects therein. Contractor shall not be liable for any loss or damage resulting from the use of materials furnished by Operator. (b) Contractor will preassemble, disassemble, or assemble materials to be furnished by Operator only when as directed by Operator and when such work can be accomplished by normal rig personnel. Contractor shall assume no liability for such service. All of such services shall be performed on a daywork basis.

18.10 Contractor's Indemnification of Operator: Contractor agrees to protect, defend, indemnify, and save Operator, its officers, directors, employees and joint owners harmless from and against all claims, demands, and causes of action of every kind and character, without limit and without regard to the cause or causes thereof or the negligence of any party or parties, arising in connection herewith in favor of Contractor's employees or Contractor's subcontractors or their employees, or Contractor's invitees, on account of bodily injury, death or damage to property. If it is judicially determined that the monetary limits of insurance required hereunder or of the indemnities voluntarily and mutually assumed under paragraph 18.10 (which Contractor and Operator hereby agree will be supported either by available liability insurance, under which the insurer has no right of subrogation against the indemnitees, or voluntarily self-insured, in part or whole) exceed the maximum limits permitted under applicable law, it is agreed that said insurance requirements or indemnities shall automatically be amended to conform to the maximum monetary limits permitted under such law.

18.11 Operator's Indemnification of Contractor: Operator agrees to protect, defend, indemnify, and save Contractor, its officers, directors, employees and joint owners harmless from and against all claims, demands, and causes of action of every kind and character, without limit and without regard to the cause or causes thereof or the negligence of any party or parties, arising in connection herewith in favor of Operator's employees or Operator's contractors or their employees or Operator's invitees other than those parties identified in paragraph 18.10 on account of bodily injury, death or damage to property. If it is judicially determined that the monetary limits of insurance required hereunder or of the indemnities voluntarily and mutually assumed under paragraph 18.11 (which Contractor and Operator hereby agree will be supported either by available liability insurance, under which the insurer has no right of subrogation against the indemnitee, or voluntarily self-insured, in part or whole) exceed the maximum limits permitted under applicable law, it is agreed that said insurance requirements or indemnities shall automatically be amended to conform to the maximum monetary limits permitted under such law.

18.12 Pollution and Contamination: Notwithstanding anything to the contrary contained herein, except the provisions of Paragraphs 15 and 18.13, it is understood and agreed by and between Contractor and Operator that the responsibility for pollution and contamination shall be as follows:

(a) Unless otherwise provided herein, Contractor shall assume all responsibility for, including control and removal of, and protect, defend and save harmless Operator from and against all claims, demands and causes of action of every kind and character arising from pollution or contamination, which originates above the surface of the land or water from spills of fuels, lubricants, motor oils, normal water base drilling fluid, pipe dope, paints, solvents, ballast, bilge and garbage, except unavoidable pollution from reserve pits, wholly in Contractor's possession and control and directly associated with Contractor's equipment and facilities.

(b) Operator shall assume all responsibility for, including control and removal of, protect, defend, indemnify and save Contractor harmless from and against all claims, demands, and causes of action of every kind and character arising directly or indirectly from all other pollution or contamination which may occur during the conduct of operations hereunder, including but not limited to, that which may result from fire, blowout, cratering, seepage or any other uncontrolled flow of oil, gas, water or other substance, as well as, the use or disposition of oil emulsion, oil base or chemically treated drilling fluids, contaminated cuttings or cavings, lost circulation and fish recovery materials and fluids.

(continued on next page)

U.S. Footage Contract continued

(c) In the event a third party commits an act or omission which results in pollution or contamination for which either Contractor or Operator, for whom such party is performing work, is held to be legally liable, the responsibility therefor shall be considered, as between Contractor and Operator, to be the same as if the party for whom the work was performed had performed the same and all the obligations respecting defense, indemnity, holding harmless and limitation of responsibility and liability, as set forth in (a) and (b) above, shall be specifically applied.

18.13 Termination of Location Liability: When Contractor has complied with all obligations of the contract regarding restoration of Operator's location, Operator shall thereafter be liable for damage to property, personal injury or death of any person which occurs as result of condition of the location and Contractor shall be relieved of such liability; provided, however, if Contractor shall subsequently reenter upon the location for any reason, including removal of the rig, any term of the contract relating to such reentry activity shall become applicable during such period.

18.14 Consequential Damages: Neither party shall be liable to the other for special, indirect or consequential damages resulting from or arising out of this Contract, including, without limitation, loss of profit or business interruptions, however same may be caused.

18.15 Indemnity Obligation: Except as otherwise expressly limited herein, it is the intent of parties hereto that all indemnity obligations and/or liabilities assumed by such parties under terms of this Contract, including without limitation, paragraphs 18.1 through 18.14 hereof, be without limit and without regard to the cause or causes thereof (including pre-existing conditions), the unseaworthiness of any vessel or vessels, strict liability, or the negligence of any party or parties, whether such negligence be sole, joint or concurrent, active or passive. The terms and provisions of paragraphs 18.1 through 18.14 shall have no application to claims or causes of action asserted against Operator or Contractor by reason of any agreement of indemnity with a person or entity not a party hereto.

19. INDEPENDENT CONTRACTOR RELATIONSHIP:

19.1 In the performance of the work herein contemplated on a "footage basis", Contractor is an independent contractor, with the authority to control and direct the performance of the details of the work, Operator being only interested in the results obtained. The work on such "footage basis" shall meet the approval of Operator and be subject to the right of inspection and supervision herein provided. Operator shall not unreasonably withhold approval of all such work, when performed by Contractor in accordance with the generally accepted practices and methods customary in the industry. Contractor agrees to comply with all laws, rules, and regulations, Federal, State, and Local, which are now, or may in the future become applicable to Contractor, Contractor's business, equipment, and personnel engaged in operations covered by this contract or accruing out of the performance of such operations; provided, however, as between Operator and Contractor specific provisions herein contained respecting the risk and responsibility for such compliance shall be controlling.

19.2 Operator shall be privileged to designate a representative or representatives who shall at all times have access to the premises for the purpose of observing tests or inspecting the work of Contractor. Such representative or representatives shall be empowered to act for Operator in all matters relating to the work herein undertaken and Contractor shall be entitled to rely on the orders and directions issued by such representative or representatives as being those of Operator.

(continued on next page)

20. NO WAIVER EXCEPT IN WRITING:

It is fully understood and agreed that none of the requirements of this contract shall be considered as waived by either party unless the same is done in writing, and then only by the persons executing this contract, or other duly authorized agent or representative of the party.

21. FORCE MAJEURE:

Neither Operator nor Contractor shall be liable to the other for any delays or damage or failure to act due, occasioned or caused by reason of any laws, rules, regulations or orders promulgated by any Federal, State or Local governmental body or the rules, regulations, or orders of any public body or official purporting to exercise authority or control respecting the operations covered hereby, including the procurance or use of tools and equipment, or due, occasioned or caused by strikes, action of the elements, water conditions, inability to obtain fuel or other critical materials or other causes beyond the control of the party affected thereby. In the event that either party hereto is rendered unable, wholly or in part, by any of these causes to carry out its obligation under this Contract, it is agreed that such party shall give notice and details of Force Majeure in writing to the other party as promptly as possible after its occurrence. In such cases, the obligations of the party giving the notice shall be suspended during the continuance of any inability so caused except that Operator shall be obligated to pay to Contractor the Work Stoppage rate set forth in Paragraph 4.3 above.

22. INFORMATION CONFIDENTIAL:

Upon written request by Operator, information obtained by Contractor in the conduct of drilling operations on this well, including, but not limited to depth, formations penetrated, the results of coring, testing, and surveying, shall be considered confidential and shall not be divulged by Contractor or his employees, to any person, firm, or corporation other than Operator's designated representative.

23. SUBCONTRACTS BY OPERATOR:

Operator may employ other contractors to perform any of the operations or services to be provided or performed by it according to Exhibit "A".

24. ASSIGNMENT:

Neither party may assign this Contract without the prior written consent of the other, and prompt notice of any such intent to assign shall be given to the other party. If any assignment is made that materially alters Contractor's financial burden, Contractor's compensation shall be adjusted to give effect to any increase or decrease in Contractor's operating costs.

25. NOTICES AND PLACE OF PAYMENT:

All notices to be given with respect to this Contract unless otherwise provided for shall be given to Contractor and to Operator respectively at the addresses hereinabove shown. All sums payable hereunder to Contractor shall be payable at his address hereinabove shown unless otherwise specified herein.

(continued on next page)

U.S. Footage Contract continued

26. SPECIAL PROVISIONS:

27. ACCEPTANCE OF CONTRACT:

The foregoing Contract is agreed to and accepted by Operator this _____________ day of _______________________________________, 19 _______

OPERATOR ____

By ____

Title ____

The foregoing Contract is accepted by the undersigned as Contractor this _____________ day of _______________________________________, 19 _______.
which is effective date of this agreement, subject to rig availability, and subject to all of its terms and provisions, with the understanding that it will not be
binding upon Operator until Operator has noted its acceptance, and with further understanding that unless said Contract is thus executed by Operator
within _____________ days of the above date Contractor shall be in no manner bound by its signature thereto.

CONTRACTOR ____

By ____

Title ____

(continued on next page)

U.S. Footage Contract continued

EXHIBIT "A"

To Drilling Contract dated ___________________________, 19 ______.

Operator ___________________________ Contractor ___________________________

Well Name and Number ___________________________

SPECIFICATIONS AND SPECIAL PROVISIONS

1. CASING PROGRAM (See Par. 7)

Size	Weight	Approx. Setting Depth	To Be Set By	Allowed Cement Time
Conductor _______ in.	_______ lbs./ft.	_______ ft.	_______	_______ hours
Surface _______ in.	_______ lbs./ft.	_______ ft.	_______	_______ hours
Protection _______ in.	_______ lbs./ft.	_______ ft.	_______	_______ hours
Oil String _______ in.	_______ lbs./ft.	_______ ft.	_______	_______ hours
Liner _______ in.	_______ lbs./ft.	_______ ft.		
Tubing _______ in.	_______ lbs./ft.	_______ ft.		

2. MUD CONTROL PROGRAM (See Par. 9)

Depth Interval (ft) From	To	Type Mud	Weight (lbs./gal.)	Viscosity (Secs)	Water Loss (cc)
_______	_______	_______	_______	_______	_______
_______	_______	_______	_______	_______	_______
_______	_______	_______	_______	_______	_______

It is understood that in the event it becomes necessary to discontinue drilling operations and to suddenly raise the mud weight _______ lbs. per gallon above the weight currently being used OR to raise the mud weight at any time to _______ lbs. per gallon, it will conclusively constitute "Abnormal Pressure" as that term is employed in Paragraph 12.2 of the Contract. Operations will thereafter go forward under the terms of such provision

(continued on next page)

U.S. Footage Contract continued

(12.2) until such condition has been overcome; the well is under control and the mud system stabilized at a weight less than _____________ lbs. per gallon, so as to permit normal drilling operations to be resumed.

Other mud specifications: ____

____

____

____

____

3. STRAIGHT HOLE SPECIFICATIONS (See Par. 9.4)

Well Depth		Maximum Distance Between Surveys, Feet	Maximum Deviation from Vertical, Degrees	Maximum Change of Angle (or Over-all Angle Between) Any Two Surveys, Degrees (1)
From	To			
________	________	________	________	________
________	________	________	________	________
________	________	________	________	________
________	________	________	________	________
________	________	________	________	________

Location of well bore at _____________ feet shall be ____________________________________

____

____

(1) a. Reduce proportionately for survey intervals less than 100 feet, but do not use intervals shorter than 30 feet.

     b. If these limits are exceeded and the distance between surveys is more than 100 feet, Contractor shall take intermediate surveys no more than 100 feet apart. If such intermediate surveys show that above limits for any interval have been exceeded, Contractor shall correct hole deviation to within limits of above specifications.

     c. When directional surveys are required, the change of angle shall be the change of over-all angle.

(continued on next page)

U.S. Footage Contract continued

4. INSURANCE (See Par. 16)

4.1 Adequate Workmen's Compensation Insurance Complying with State Laws applicable or Employers' Liability Insurance with limits of $_________________ covering all of Contractor's employees working under this agreement.

4.2 Comprehensive Public Liability Insurance or Public Liability Insurance with limits of $_________________ for the death or injury of any one person and $_________________ for each accident.

4.3 Comprehensive Public Liability Property Damage Insurance or Public Liability Property Damage Insurance with limits of $_________________ for each accident and $_________________ aggregate per policy.

4.4 Automobile Public Liability Insurance with limits of $_________________ for the death or injury of each person and $_________________ for each accident; and Automobile Public Liability Property Damage Insurance with limits of $_________________ for each accident.

4.5 In the event operations are over water, Contractor shall carry in addition to the Statutory Workmen's Compensation Insurance, endorsements covering liability under the Longshoremen's & Harbor Workers' Compensation Act and Maritime liability including maintenance and cure with limits of $_________________ for each death or injury to one person and $_________________ for any one accident.

4.6 Other Insurance: ____

____

____

____

5. EQUIPMENT, MATERIALS AND SERVICES TO BE FURNISHED BY CONTRACTOR:

The machinery, equipment, tools, materials, supplies, instruments, services and labor hereinafter listed, including any transportation required for such items, shall be provided at the location at the expense of Contractor unless otherwise noted hereon.

5.1 Drilling Rig:

Complete drilling rig, designated by Contractor as his Rig No. ________________________________, the major items of **equipment** being:

Drawworks:

(MAKE AND MODEL)

Engines: Make, Model, and H.P. ____

No. on Rig ____

(continued on next page)

U.S. Footage Contract continued

Pumps: No. 1 Make, Size, and Power ____

No. 2 Make, Size, and Power ____

Mud Mixing Pump: Make, Size, and Power ____

Boilers: Number, Make, H.P. and W.P. ____

Derrick or Mast: Make, Size, and Capacity ____

____

Substructure: Size and Capacity ____

Rotary Drive: Type ____

Drill Pipe: Size ____________ in. ____________ ft.; Size ____________ in. ____________ ft.

Drill Collars: Number and Size ____

Blowout Preventers: ____

Size	Series or Test Pr.	Make & Model	Number
__________	__________	__________	__________
__________	__________	__________	__________
__________	__________	__________	__________

B.O.P. Closing Unit: __________ __________ __________

B.O.P. Accumulator: __________ __________ __________

5.2 Trucking service and other transportation, hauling, or winching services as required to move Contractor's property to location, rig up Contractor's rig, tear down Contractor's rig, and remove all of Contractor's property from location.

5.3 Drilling bits, reamers, stabilizers, reamer cutters, and other drilling tools or devices (except while on daywork).

5.4 Contract fishing tool services and fishing tool rentals (except while on daywork).

5.5 Derrick timbers.

5.6 Normal strings of drill pipe and drill collars specified above.

5.7 Conventional drift indicator.

5.8 Circulating mud pits.

5.9 Necessary pipe racks and rigging up material.

5.10 Normal storage for mud and chemicals.

5.11 Shale Shaker.

(continued on next page)

6. EQUIPMENT, MATERIALS AND SERVICES TO BE FURNISHED BY OPERATOR:

The machinery, equipment, tools, materials, supplies, instruments, services and labor hereinafter listed, including any transportation required for such items, shall be provided at the location at the expense of Operator unless otherwise noted hereon.

6.1 Furnish and maintain adequate roadway and/or canal to location, right-of-way, including rights-of-way for fuel and water lines, river crossings, highway crossings, gates and cattle guards.

6.2 Stake location, clear and grade location, and provide turnaround, including surfacing when necessary.

6.3 Test tanks with pipe and fittings.

6.4 Mud storage tanks with pipe and fittings.

6.5 Separator with pipe and fittings.

6.6 Labor to connect and disconnect mud tank, test tank, and separator.

6.7 Labor to disconnect and clean test tanks and separator.

6.8 Drilling mud, chemicals, lost circulation materials and other additives.

6.9 Pipe and connections for oil circulating lines.

6.10 Labor to lay, bury and recover oil circulating lines.

6.11 Drilling bits, reamers, reamer cutters, stabilizers and special tools while operating on daywork basis.

6.12 Contract fishing tool services and tool rental while operating on a daywork basis.

6.13 Wire line core bits or heads and wire line core catchers if required.

6.14 Conventional core bits and core catchers.

6.15 Diamond core barrel with head.

6.16 Cement and cementing service.

6.17 Electrical and Gamma-Neutron and Micro logging services.

6.18 Directional, caliper, or other special services.

6.19 Gun or jet perforating services.

6.20 Explosives and shooting devices.

6.21 Formation testing, hydraulic fracturing, acidizing and other related services.

6.22 Equipment for drill stem testing.

6.23 Mud logging services.

6.24 Sidewall coring service.

6.25 Welding service for welding bottom joints of casing, guide shoe, float shoe, float collar and in connection with installing of well head equipment if required.

6.26 Casing, tubing, liners, screen, float collars, guide and float shoes and associated equipment.

6.27 Casing scratchers and centralizers.

(continued on next page)

U.S. Footage Contract continued

6.28 Well head connections and all equipment to be installed in or on well or on the premises for use in connection with testing, completion and operation of well.

6.29 Special or added storage for mud and chemicals.

6.30 Casinghead, API series, to conform to that shown for the blowout preventers specified in Paragraph 5.1 above.

6.31 Blowout Preventer testing packoff.

6.32 Casing Thread Protectors and Casing Lubricant.

7. EQUIPMENT, MATERIALS AND SERVICES TO BE FURNISHED BY DESIGNATED PARTY:

The machinery, equipment, tools, materials, supplies, instruments, services, and labor listed as the following numbered items including any transportation required for such items unless otherwise specified, shall be provided at the location and at the expense of the party hereto as designated by an X mark in the appropriate column.

Item	To Be Provided By and At The Expense Of	
	Operator	Contractor
7.1 Cellar and runways	__________	__________
7.2 Fuel (located at ______________)	__________	__________
7.3 Fuel Lines (length ______________)	__________	__________
7.4 Water at source including required permits	__________	__________
7.5 Water well including required permits	__________	__________
7.6 Water lines including required permits	__________	__________
7.7 Water storage tanks __________ capacity	__________	__________
7.8 Labor to operate water well or water pump	__________	__________
7.9 Maintenance of water well, if required	__________	__________
7.10 Mats for engines and boilers, or motors and mud pumps	__________	__________
7.11 Transportation of Contractor's property:		
Move in	__________	__________
Move out	__________	__________
7.12 Materials for "boxing in" rig and derrick	__________	__________

(continued on next page)

U.S. Footage Contract continued

7.13 Special strings of drill pipe and drill collars as follows:

7.14 Kelly joints, subs, elevators and slips for use with special drill pipe

7.15 Drill pipe protectors for Kelly joint and each joint of drill pipe running inside of Surface Casing as required, for use with normal strings of drill pipe

7.16 Drill pipe protectors for Kelly joint and drill pipe running inside of Protection Casing

7.17 Coring reel with wire line of sufficient length for coring at maximum depth specified in contract

7.18 Wire line core barrel

7.19 Conventional core barrel

7.20 Rate of penetration recording device

7.21 Extra labor for running and cementing casing (Casing Crews)

7.22 Casing tools

7.23 Rig time for running of casing-conductor

7.24 Rig time for running of casing-surface

7.25 Rig time for running of casing protection

7.26 Rig time for running of casing production

7.27 Rig time for running of casing liner

7.28 Power casing tongs

7.29 Tubing tools

7.30 Power tubing tong

7.31 Swabbing unit with swabbing line

7.32 Drilling mouse and rat holes

7.33 Drilling hole for or driving for conductor pipe

7.34 Reserve pits

(continued on next page)

U.S. Footage Contract continued

7.35 Erect and dismantle derrick .... ___________ ___________
7.36 Special Foundation .... ___________ ___________
7.37 Swab .... ___________ ___________
7.38 Swab lubricator .... ___________ ___________
7.39 Swab rubbers .... ___________ ___________
7.40 Crew Boats, Number ___________________ .... ___________ ___________
7.41 Service Barge .... ___________ ___________
7.42 Service Tug Boat .... ___________ ___________
7.43 Helicopter Service .... ___________ ___________
7.44 Upper Kelly Cock .... ___________ ___________
7.45 Camp Facilities _______________ men .... ___________ ___________
7.46 Catering .... ___________ ___________
7.47 Charges, cost of bonds for public roads .... ___________ ___________
7.48 ____ ___________ ___________
7.49 ____ ___________ ___________
7.50 ____ ___________ ___________

OTHER PROVISIONS:

Signed by the
Parties as correct:

For Contractor ____

For Operator ____

(continued on next page)

U.S. Footage Contract continued

EXHIBIT "B"
(See Paragraph 9.5)

The following clauses, where required by law, are incorporated in the Contract by reference as if fully set out:

(1) The Equal Opportunity Clause prescribed in 41 CFR 60-1.4.

(2) The Affirmative Action Clause prescribed in 41 CFR 60-250.4 regarding veterans and veterans of the Vietnam era.

(3) The Affirmative Action Clause for handicapped workers prescribed in 41 CFR 60-741.4.

(4) The Certification of Compliance With Environmental Laws prescribed in 40 CFR 15.20.

Appendix B
Common Oilwell Drilling Calculations

Oilfield calculations are simple to understand and easy to apply to the drilling of the well. This appendix will explain how to calculate the following:

1. Bottom hole pressure (BHP)
2. Capacities of tubular products
3. Annular capacities of tubular products in a wellbore
4. Cement requirements
5. Drill collar weight needed for weight on bit
6. Equivalent mud weight for shoe testing
7. Surface-to-bit calculations
8. Bit-to-surface calculations
9. Number of pump strokes vs. barrels of fluid
10. Kill mud needed
11. Final circulating pressure

Bottom Hole Pressure

BHP = mud weight (ppg) × 0.052 (constant) × depth

Example

$$\text{Well} = 10,000 \text{ ft deep}$$
$$\text{Mud weight} = 11 \text{ ppg}$$
$$\text{Constant} = 0.052$$

11 ppg × 0.052 × 10,000 ft = __________

So the BHP at 10,000 ft is 5,720 psi.

Capacity of Tubular Products

The capacity of tubular products—casing, drill pipe, and tubing—can be found in the capacity section of your cement book. The capacities come in the following categories:

- Cubic feet per linear foot
- Linear feet per cubic foot
- Barrels per linear foot
- Linear feet per barrel
- Gallons per linear foot
- Linear feet per gallon

Example

Barrels per linear foot:

	Barrels/Foot
Drill pipe 4.5 in. 16.60 lb/ft	0.01422
Casing 7 in. 23 lb/ft	0.03940
Tubing 2.85 in. 9.50 lb/ft	0.00468

To get the barrels simply multiply the depth by the barrels per linear foot.

Example

In a 3,000 ft well using 4.5-in. 16.60 lb/ft drill pipe use the following:

0.01422 bbl/ft $\times$ 3,000 ft = ________

the barrels in 3,000 ft of drill pipe.

Annular Capacity of Tubular Products in a Wellbore

The annular capacity can be found under the "annular volume between tubing and casing or open hole" section in your cement book. This includes:

- Drill pipe and open hole
- Drill pipe and casing
- Casing and open hole
- Casing and casing
- Tubing and casing
- Tubing and open hole

Example

The cement book gives you the capacity for the area between 2⅜-in. tubing and 4.5-in. 9.50 lb/ft casing as 0.0108 bbl/ft. To get the total capacity of the area multiply the length of tubing by 0.0108 bbl/ft:

0.0108 bbl/ft $\times$ 3,000 ft = ________

The total capacity is 32.4 bbl.
Other annular capacities are calculated similarly.

Cement Calculations

- Salt (NaCl) percentages are always calculated based on the weight of the mix water.
- All other admix percentage calculations are based on the weight of the cement.
- All soluble materials less than 5% of the weight of cement are ignored in the calculations.
- All insoluble additives are included in the calculations.
- Water always weighs 8.33 lb/gal.
- One cubic foot is equal to 7.48 gal.
- Always add a minimum 5.2% mix of water for each 1% of bentonite added to a cement system.

The following calculations will help you do cement jobs more easily. Until you learn shortcuts, stay with each procedure.

Slurry Calculations

To find the slurry density (the weight per volume measured in lb/gal of a cement slurry):

Component's weight (lb) $\times$ Absolute volume (gal/lb) = Volume (gal)

$$\text{slurry density} = \frac{\text{Total components' weight (lb)}}{\text{Total components' volume (gal)}}$$

Example

Class "H" cement weighs 94 lb; absolute volume 0.0382 water (46%) weighs 43.24 lb; absolute volume 0.12

$$94 \text{ lb} \times 0.0382 = 3.59 \text{ gal}$$

$$43.24 \text{ lb} \times 0.12 = 5.19 \text{ gal}$$

$$3.59 \text{ gal} + 5.19 \text{ gal} = 8.78 \text{ total components' volume}$$

$$94 \text{ lb} + 43.24 \text{ lb} = 137.24 \text{ total components' weight}$$

$$\text{slurry density} = \frac{137.24 \text{ lb}}{8.78 \text{ gal}}$$

$$= 15.6 \text{ lb/gal}$$

To find the slurry yield (the volume or cubic feet of one sack of cement after adding water and admixes):

$$\text{slurry yield} = \frac{\text{total components' volume (gal)}}{7.48 \text{ gal/ft}^3}$$

$$\text{slurry yield} = \frac{8.78 \text{ gal}}{7.48 \text{ gal/ft}^3} = 1.17 \text{ ft}^3/\text{sk}$$

To find the total number of sacks of cement needed:

$$\text{sacks of cement} = \frac{\text{annular volume (ft}^3)}{\text{slurry yield (ft}^3/\text{sk})}$$

To convert the total volume into barrels:

$$\text{slurry volume (bbl)} = \text{slurry volume (ft}^3) \times 0.1781$$
(constant)

Drill Collar Weight

To determine how many drill collars are needed to apply weight to the bit, first decide how much weight you want to run on the bit and multiply that amount by 0.25 for a

25% increase in weight. Then determine the weight in pounds per feet of the drill collars you are using, and divide the total number of pounds by the pounds per feet. Finally, divide the total number of feet by the average length of the drill collars.

Example

You want to run 30,000 lb of weight on the drill bit, and the drill collars you have weigh 75 lb/ft.

$$30,000 \text{ lb} \times 0.25 = 7,500 \text{ lb}$$

$$30,000 \text{ lb} + 7,500 \text{ lb} = 37,500 \text{ lb}$$

$$\frac{37,500 \text{ lb}}{75 \text{ lb/ft}} = \begin{array}{l} 500 \text{ ft of drill collars needed} \\ \text{to obtain the specified weight} \end{array}$$

If the drill collars have an average length of 30 ft:

$$\frac{500 \text{ ft}}{30 \text{ ft}} = \begin{array}{l} 16.66 \text{ drill collars needed in the BHA} \\ (17 \text{ DCs rounded off}) \end{array}$$

Equivalent Mud Weight for Shoe Testing

To find the EMW for shoe testing purposes, you must first determine the bottom hole pressure (BHP), by multiplying the mud weight by the depth by the constant 0.052.

Example

With an 8.9 ppg mud in hole at a depth of 3,014 ft use the formula for bottom hole pressure:

$$8.9 \text{ ppg} \times 0.052 \times 3,014 \text{ ft} = \begin{array}{l} 1,394.87 \text{ psi bottom} \\ \text{hole pressure} \end{array}$$

Now find out what is needed for your EMW. If for example, it is 13.5 ppg, take that figure and find BHP:

13.5 ppg $\times$ 0.052 $\times$ 3,014 ft = 2,115.82 psi

Now take the figures and subtract:

    2,116 psi rounded off
  (1,395) psi rounded off
    ——————
     721 psi

which leaves the 721 psi. Using a pumping unit or the rig pump, get ready to build to 721 psi. First close your annular preventer to shut the well in, then pump up to 721 psi, using 100-lb increments. When the pressure reaches 721 psi on the pump, let it sit for 15 minutes and run a bleed-off test. If there is a loss of more than 15% then a squeeze job is necessary. If not, you can go to drilling.

If it bleeds off, first look for surface leaks. If there are none, then the problem is with the formation. For example, if it bleeds down to 300 psi, you repump to 720 psi, and it again bleeds down, record the pressure and determine the maximum mud weight the shoe will hold. Use the following method:

1,395 psi + 300 psi $\div$ 0.052 $\div$ 3,014 ft = 10.81 ppg

Thus, the well tested will hold only a 10.81 mud weight, and it will be necessary to squeeze the formation.

Surface-to-Bit Calculations

For surface-to-bit calculations, first find the capacity of the drill pipe (DP) and the drill collars (DC) and divide that by barrels per stroke times strokes per minute.

Example

$$S \text{ to } B = \frac{\begin{array}{c}(0.1422 \text{ bbl/ft DP})(7,408 \text{ ft}) \\ + (0.01776 \text{ bbl/ft DC})(592)\end{array}}{(0.09 \text{ bbl/stroke})(60 \text{ strokes/minute})}$$

$$S \text{ to } B = \frac{105.34 \text{ barrels} + 10.51 \text{ barrels}}{5.4}$$

$$S \text{ to } B = \frac{115.85}{5.4}$$

$$S \text{ to } B = 21.45 \text{ min}$$

Bit-to-Surface Time

The equation for this is the same as for the surface-to-bit calculation, except that the drill pipe and drill collar capacities are found for the annular side (OH = open hole).

$$B \text{ to } S = [(\text{bbl/ft between DP and OH})(\text{depth}) + (\text{bbl/ft between DC and OH})(\text{ft of DC})]/[(\text{bbl/stroke})(\text{stroke/min})]$$

Pump Stroke Calculations

To find the number of pump strokes needed to pump a certain amount of fluid divide the number of barrels of fluid to be pumped by the barrels per stroke.

Example

$$\text{Strokes} = \frac{\text{bbl}}{\text{bbl/stroke}}$$

Kill Mud Calculations

Using the simple equation:

$$W2 = \frac{SIDPP}{0.052 \times depth} + W1$$

W1 = original mud weight

W2 = kill weight

SIDPP = shut-in drill pressure

Appendix C
The Cement Book

This reference is provided by cement suppliers and is called by various names by various companies, such as the *Cementers Field Book* or the *Engineer's Handbook*. It is referred to in this book simply as the cement book. This book is a valuable tool to the field engineer as the tables eliminate the need for lengthy calculations. With the cement book a well can be drilled from start to finish with a minimum of mathematical effort. If you can master the use of this reference, you can match skills with the engineers back in the office or service companies in the field. All of the data on drill pipe, casing, and open holes has been worked out by professionals and tabulated in the cement book.

Capacity

The cement book will help you determine the capacity of tubing, casing, and drill pipe. Capacity in this case is the inside area of tubular products or the inside area of an open hole without casing or drill pipe in the hole.

For 4.5-in. 16.60 lb/ft drill pipe, locate OD size 4.5 in. and then find 16.60 lb/ft with tool joints. Under the column marked ''barrels per linear foot will be the figure, 0.01422. That figure means that in one foot of 4.5 in. 16.60 lb/ft drill pipe there is a 0.01422 barrel capacity. To find the total capacity of a given length of drill pipe, simply multiply the number 0.01422 times the length.

1,000 × 0.01422 = 14.22 barrels (see Figure C-1A)

So 1,000 ft of 4.5-in. 16.60 lb/ft drill pipe would have a volume of 14.22 bbl.

Example

For 4.5-in. 10.50 lb/ft casing, look in the cement book and locate 4.5-in. OD casing weighing 10.50 lb/ft with coupling. Go across the page to the column marked ''barrels per linear foot'' and find the number 0.0159. To calculate the volume in 1,000 ft of this casing:

1,000 × 0.0159 = 15.9 bbl

As you can see, using the tables in the cement book is simple.

Annular Volume and Height

The cement book will help you find the annular volumes between drill pipe and casing or open hole. This section of the book covers small casing, tubing, and drill pipe.

The annular space is the space found between two cylinders. In the case of drilling, it is the space between tubing and casing, tubing and open hole, casing and casing, casing and open hole, drill pipe and casing, or drill pipe and open hole. In engineering handbooks, the annular space is referred to as the annular volume.

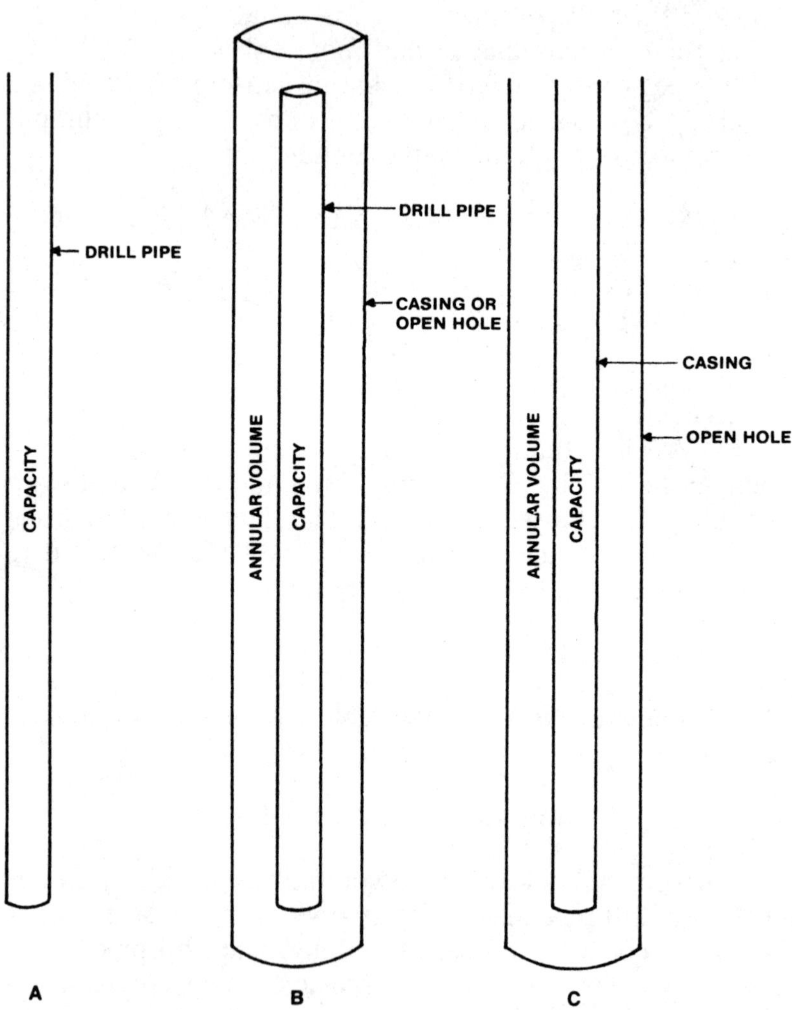

Figure C-1. The information provided in the cement book.

Example

Find the annular volume for a 4.5-in. 16.60 lb/ft drill pipe in 7⅞-in. diameter 1,000-ft long open hole.

Look in the cement book and find 4.5-in. OD drill pipe. Then look for 7⅞-in. open hole (OH). Looking under barrels per linear foot we find 0.0406 bbl/ft. Using the simple calculation:

$$1,000 \text{ ft} \times 0.0406 \text{ bbl/ft} = 40.6 \text{ bbl}$$

So the volume for the dimensions given to us is 40.6 bbl.

To find the total volume of the drill pipe and the annulus, simply combine what we already know (see Figure C-1B).

$$\text{(capacity of DP)} — 1,000 \times 0.01422 = 14.22 \text{ bbl}$$
$$\text{(annular capacity)} — 1,000 \times 0.0406 = \underline{40.60} \text{ bbl}$$
$$\text{Total capacity of hole with pipe} = 54.82 \text{ bbl}$$

The annular volume between casing and casing, or casing and open hole including surface casing and longer strings.

Example

For a 7-in. casing in a 9⅞-in. diameter 1,000 ft deep open hole, look up 9⅞-in. open hole under ''barrels per linear foot'' to find the figure 0.0471. Multiply by 1,000 ft:

$$1,000 \text{ ft} \times 0.0471 = 47.1 \text{ bbl/ft}$$

To find the annular volume in cubic feet, look in the cement book for a 9⅞-in. open hole under the column ''cubic feet per linear foot,'' and find the number 0.2647. To find the total volume in cubic feet, multiply 0.2647 by 1,000 ft:

$$1{,}000 \text{ ft} \times 0.2647 = 264.7 \text{ ft}^3 \text{ (see Figure C-1C)}.$$

Appendix D
The Consultant's Checklist

This appendix contains all the forms you need for a complete record of the drilling operation. If you keep all these records, you will stay out of trouble and have everything covered.

Turn in the following at the end of the job:

1. The daily reports (consultant's reports)
2. The daily mud reports
3. The geolograph reports
4. The pit indicator and flow shows reports
5. Invoices for services
6. Rental tool receiving slips
7. Rental tool pick-up slips
8. Daily rig report
9. Bit records
10. The mud logger's report

The Daily Report covers the day-to-day operation of the well. It should include:

1. Well number and lease number
2. Operator's name
3. Mud evaluation daily information
4. Drilling costs, daily and accumulated, for 24 hours of operation, broken down in detail
5. The bottom hole assembly and length
6. Depth of well and footage drilled in 24 hours

The daily drilling report is a good way to look back at the well program and evaluate what happened at a certain depth.

Daily Mud Reports give mud data on a day-to-day basis as well as daily and cumulative costs.

The Geolograph records the weight on the string, downtime, and footage. The record will be left in a book on the rig floor. Collect all the sheets and turn them in.

Pit Indicator and Flow Shows. The tape needs to be turned in. Simply roll it up and add to the box.

Typical Drilling Operation Forms

BOP PRESSURE TESTING

Check List (circle one)

	Pounds	Pounds	Pounds
BOP Tested	5,000	3,000	2,500
Annular Preventer Tested	2,500	2,000	1,500
HCR Valve Tested	5,000	3,000	2,500
Choke Lines Tested	5,000	3,000	2,500
Superchoke Tested	5,000	3,000	2,500
Kill Line Tested	5,000	3,000	2,500
TIW Valve Tested	5,000	3,000	2,500

Bottom Hole Assembly Check List

	Size	Manufacturer	Rental Fee
Shock Sub	__________	__________	__________
Stabilizer	__________	__________	__________
Drilling Jar	__________	__________	__________
Roller Reamer	__________	__________	__________
Collar Size	__________	__________	__________
Bit Size	__________	__________	__________

Maximum Weight Needed on Bit ____

(continued on next page)

CORE SAMPLES, DRILL STEM TESTS, LOGS

Core Sample: Depth and Tools Used

	Depth	Tool Used & Service Company	Cost
CS 1.	_______	_________________________	_______
CS 2.	_______	_________________________	_______
CS 3.	_______	_________________________	_______
CS 4.	_______	_________________________	_______
CS 5.	_______	_________________________	_______

Drill Stem Tests

DST 1.	_______	_________________________	_______
DST 2.	_______	_________________________	_______
DST 3.	_______	_________________________	_______
DST 4.	_______	_________________________	_______
DST 5.	_______	_________________________	_______
DST 6.	_______	_________________________	_______
DST 7.	_______	_________________________	_______

Logs

Services Used _______________________________________

____

T.D. _______________________ To _______________________

Production Zones _______ _______ _______

_______ _______ _______

Cost _______________________________________

(continued on next page)

NUMBER OF DRILL COLLARS NEEDED

Step 1

Maximum Weight on Bit	25% Buoyant Factor	Added Weight Needed	Maximum Weight on Bit	Collar Weight Needed to Drill the Hole

________ x ________ 0.25 = ________ + ________ = ________________

Step 2

Collar   Footage
Weight   Needed
Needed

________ = ________
Wt/Ft

Step 3

Footage   Number of
Needed   Collars Needed   Rounded Off to Next Collar

________ = ________________ = ________________________
(Average
Length of
Collars is
30 ft.)

Total Collars Needed   ________________________________

Number Stabilizers Needed .________________________________

Location of Stabilizers   ________________________________

What Combination (Example: 30, 90, 120) ______ ______ ______

        Set in String   ______ ______ ______

INTERMEDIATE STRING AND LONG STRING

Casing Company ____

Cement Company ____

(continued on next page)

No. Joints Used ________________   No. Joints Unused ____________

Length of Casing ____

No. Barrels Cement Pumped _____________________________________

No. Barrels Displacement Fluid __________________________________

Cost of Operation ____________________   WOC _________________

Notes: ____

____

____

No. Sacks Cement Needed ______________   Size of Hole _____________

Type Cement and Additives _____________________________________

Casing Size __________________   Weight ____________________

Location of Centralizers ______________________________________

Location of Float Collar ______________________________________

NIPPLE UP

Casing Head Size ______________   Manufacturer _________________

Welded By ____

Tested By ____

Location of Casing Flange to Ground Level _________________________

Rig Elevation ____

BOP DRAWINGS

(Draw location of rams and their sizes)

(continued on next page)

Manual Closing Devices Installed
_________   _________
yes   no

Accumulator Tested and Charged
_________   _________
yes   no

All Accumulator Leaks Repaired
_________   _________
yes   no

RENTAL EQUIPMENT

	Item	Rental Fee	Arrived	Shipped	Est. Cost
1.	_______	_______	_______	_______	_______
2.	_______	_______	_______	_______	_______
3.	_______	_______	_______	_______	_______
4.	_______	_______	_______	_______	_______
5.	_______	_______	_______	_______	_______

CHECK LIST

Location Name and Lease No. _______________________________________

Drilling Contractor ____

Emergency Phone No. ____

Toolpusher Name(s) ____

SURFACE CASING

Casing Crew Company ____

Hammer Company ____

(continued on next page)

Cement Company ____

Number Sacks Needed ____

Type Cement and Additives ______________________________________

Hole Size __________   Casing Size __________   Casing Weight ______

Bit Manufacturer ____

Type Guide Shoe and Float Collar ________________________________

Location of FC and Centralizers __________________________________

____

Number and Footage of Joints Returned ___________________________

____

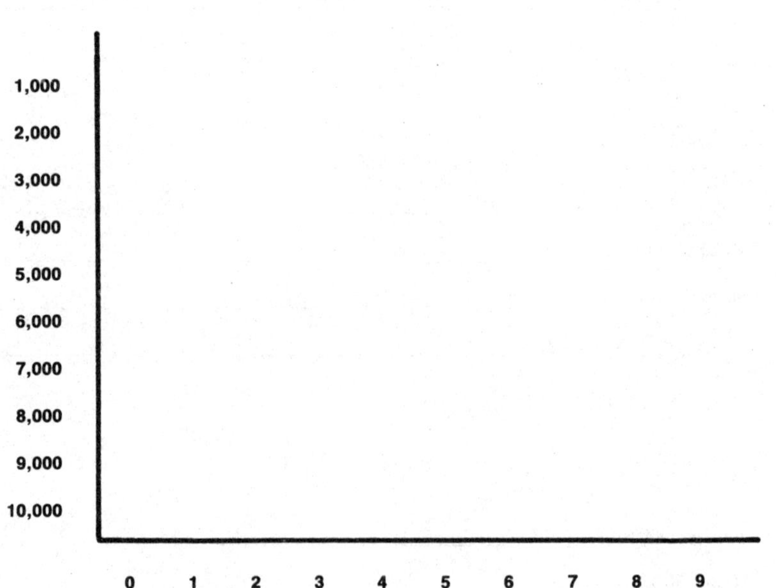

DRIFT CHART

BOTTOM HOLE ASSEMBLY CHART

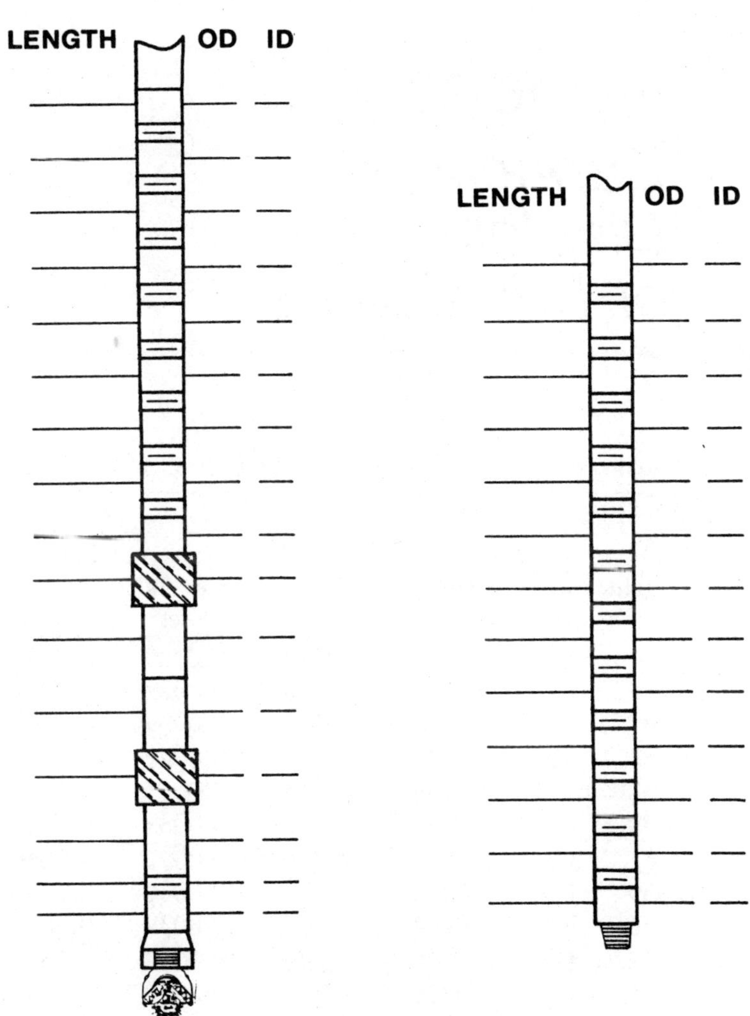

Appendix E
Capacity and Displacement of Drill Collars

Capacity and Displacement of Drill Collars

Drill Collar			Displacement		Capacity	
OD Size (in.)	Weight (lbs/ft)	ID Size (in.)	Barrels/ Feet	Barrels/ 93-feet Stand	Barrels/ Feet	Barrels/ 93-feet Stand
$4^1/_2$	51	1	0.0187	1.740	0.0009	0.0837
	58	$1^1/_2$	0.0175	1.620	0.0022	0.2046
	43	2	0.0158	1.470	0.0039	0.3627
$4^3/_4$	54	$1^1/_2$	0.0197	1.830	0.0022	0.2046
	52	$1^3/_4$	0.0189	1.757	0.0030	0.2790
	50	2	0.0180	1.670	0.0039	0.3627
5	61	$1^1/_2$	0.0221	2.055	0.0022	0.2046
	59	$1^3/_4$	0.0213	1.980	0.0030	0.2790
	56	2	0.0204	1.890	0.0039	0.3627
$5^1/_4$	68	$1^1/_2$	0.0246	2.290	0.0022	0.2046
	65	$1^3/_4$	0.0238	2.210	0.0030	0.2790
	63	2	0.0229	2.130	0.0039	0.3627
$5^1/_2$	75	$1^1/_2$	0.0272	2.530	0.0022	0.2046
	73	$1^3/_4$	0.0264	2.460	0.0030	0.2790
	70	2	0.0255	2.370	0.0039	0.3627

(continued on next page)

Appendix E continued

Drill Collar			Displacement		Capacity	
OD Size (in.)	Weight (lbs/ft)	ID Size (in.)	Barrels/ Feet	Barrels/ 93-feet Stand	Barrels/ Feet	Barrels/ 93-feet Stand
$5^3/_4$	82	$1^1/_2$	0.0299	2.780	0.0022	0.2046
	80	$1^3/_4$	0.0291	2.710	0.0030	0.2790
	78	2	0.0282	2.620	0.0039	0.3627
6	88	$1^3/_4$	0.0320	2.980	0.0030	0.2790
	85	2	0.0311	2.890	0.0039	0.3627
	83	$2^1/_4$	0.0301	2.800	0.0049	0.4557
$6^1/_4$	96	$1^3/_4$	0.0349	3.240	0.0030	0.2790
	94	2	0.0340	3.160	0.0039	0.3627
	91	$2^1/_4$	0.0330	3.070	0.0049	0.4557
$6^1/_2$	105	$1^3/_4$	0.0380	3.530	0.0030	0.2790
	102	2	0.3710	3.450	0.0039	0.3627
	99	$2^1/_4$	0.0361	3.360	0.0049	0.4557
$6^3/_4$	114	$1^3/_4$	0.0413	3.840	0.0030	0.2790
	111	2	0.0404	3.760	0.0039	0.3627
	108	$2^1/_4$	0.0394	3.660	0.0049	0.4557
7	120	2	0.0437	4.060	0.0039	0.3627
	114	$2^1/_2$	0.0415	3.860	0.0061	0.5673
	107	3	0.0388	3.610	0.0088	0.8184
$7^1/_4$	130	2	0.0472	4.390	0.0039	0.3627
	124	$2^1/_2$	0.0450	4.180	0.0061	0.5673
	116	3	0.0423	3.930	0.0088	0.8184
$7^1/_2$	139	2	0.0507	4.720	0.0039	0.3627
	133	$2^1/_2$	0.0485	4.510	0.0061	0.5673
	126	3	0.0458	4.260	0.0088	0.8184
$7^3/_4$	144	$2^1/_2$	0.0522	4.850	0.0061	0.5673
	136	3	0.0495	4.600	0.0088	0.8184
	128	$3^1/_2$	0.0464	4.320	0.0119	1.1070
8	147	3	0.0534	4.970	0.0088	0.8184
	143	$3^1/_4$	0.0519	4.830	0.0103	0.9580
	138	$3^1/_2$	0.0503	4.680	0.0119	1.1070

Appendix F
*Pump Output Table**

*Courtesy of IMCO Services

Table 1

DISPLACEMENT OF DUPLEX MUD PUMPS
(90% Efficiency)

Displacement expressed in barrels per stroke (bps)

Liner Size	STROKE LENGTH (in.)								
(in.)	8	10	12	14	15	16	18	20	22
4	.034	.044	.053	.062					
4¼	.040	.050	.060	.070					
4½	.044	.056	.067	.078	.084	.087	.090	.099	
4¾	.050	.062	.075	.087	.093	.096	.102	.113	
5	.055	.069	.083	.096	.103	.110	.115	.138	
5¼	.061	.076	.091	.106	.114	.121	.137	.152	
5½	.067	.083	.099	.115	.121	.127	.142	.167	
5¾	.073	.091	.109	.123	.137	.146	.164	.182	
6	.079	.099	.119	.139	.146	.153	.172	.198	
6¼	.086	.108	.130	.151	.161	.168	.189	.215	
6½	.093	.116	.141	.164	.176	.182	.205	.228	
6¾	.100	.126	.152	.178	.188	.198	.223	.247	
7	.108	.135	.164	.192	.202	.214	.241	.267	
7¼	.115	.145	.174	.201	.217	.231	.259	.288	
7½		.155	.186	.217	.232	.248	.279	.310	.342
7¾			.199	.232	.248	.265	.299	.331	.365
8		.176	.212	.247	.264	.282	.317	.355	.413
8½			.225	.263	.281	.300	.338	.375	.439

Table 2

DISPLACEMENT OF TRIPLEX MUD PUMPS
(100% Efficiency)

Displacement expressed in barrels per stroke (bps)

Liner Size	STROKE LENGTH (in.)								
(in.)	7	7½	8	8½	9	9¼	10	11	12
3	.015	.016	.017	.019	.020	.020	.022	.024	.026
3¼	.018	.019	.021	.022	.023	.024	.026	.028	.031
3½	.021	.022	.024	.025	.027	.028	.030	.033	.036
3¾	.024	.026	.027	.029	.031	.032	.034	.038	.041
4	.027	.029	.031	.033	.035	.036	.039	.043	.047
4¼	.031	.033	.035	.037	.039	.041	.044	.048	.053
4½	.034	.037	.039	.042	.044	.045	.049	.054	.059
4¾	.038	.041	.044	.047	.049	.051	.055	.060	.066
5	.043	.045	.049	.052	.055	.056	.061	.067	.073
5¼	.047	.050	.054	.057	.060	.062	.067	.074	.080
5½	.051	.055	.059	.062	.066	.068	.073	.081	.088
5¾	.056	.060	.064	.068	.072	.074	.080	.088	.096
6	.061	.065	.070	.074	.079	.081	.087	.096	.105
6¼	.066	.071	.076	.081	.085	.088	.095	.104	.114
6½	.072	.077	.082	.087	.092	.095	.103	.113	.123
6¾	.077	.083	.088	.094	.100	.102	.111	.122	.133
7	.083	.089	.095	.101	.107	.110	.119	.131	.143

Glossary

Abstract company—Company that prepares abstracts of title.

Abstract of title—Book or file containing a summary of every document affecting title to a property.

AFE—Authorized field expenditures. This is a document filled out by the engineer that estimates the cost of drilling the well, so approval for the necessary funds can be obtained.

Air drilling—A type of drilling that uses air or gas to move cuttings. It is 30%–40% faster than mud drilling.

Anchor deadline—Device to hold the deadline to the derrick or substructure. It is also the main element of the weight indicator.

Annular preventer—A blowout preventer that closes around the pipe.

Annular space—The open space between the casing and the open hole or the casing and the casing or the tubing and the casing or the tubing and the open hole.

Automatic driller—Device that controls the weight on the bit, allowing the driller to be free of the brake.

Backside—The annular space. To pump the backside is to pump fluid down the annular space.

Bales—Heavy metal devices used to hook elevators to the traveling block.

Barite—Substance used to weight up (increase the density of) drilling mud. It increases the specific gravity of the mud mix.

Barrel—American Petroleum Institute (API) measure equal to 42 U.S. gallons.

Bell nipple—Piece of pipe installed at the top of the blowout preventer stack to return drilling fluid to the mud pits through the flow line and to guide the first few feet of pipe in the hole.

Bentonite—Substance added to drilling fluid to increase viscosity or mud thickness.

BHA—See *Bottom hole assembly*.

Bit—The cutting tool in rotary drilling. It comes in three designs: with roller bearings (sealed and journal), diamond bits, and drag bits.

Bit box—A locked box used to store bits on location.

Bit breaker—Tool to breakout or make up bits; similar to a wrench.

Bit jet—An insert to control hydraulics in a bit for mud flow return.

Bit sub—The sub that joins drill collars to the drill bit or the shock sub to the drill collars.

Blind rams—Hydraulic-driven closure device used to seal the well when no drill pipe is in the hole.

Block—A device with pulleys for handling the weight of the drill string.

Blowout—The uncontrolled escape of oil or gas from a well.

Blowout preventer (BOP)—Manual- or hydraulic-operated device used above the casing to seal off fluids coming up the annular space. Blind rams seal the hole when no pipe is in the hole. Pipe rams seal the hole when pipe is in the hole. Annular BOPs allow the drill pipe to move so it does not get stuck.

Bonus—Cash payment given to the landowner at the time he signs the oil and gas lease.

BOP—See *Blowout preventer*.

Bottom hole assembly (BHA)—Everything below the drill pipe in the drill stem.

Bradenhead—Also called the casing or wellhead, it is welded to the casing to which the BOP is attached.

Breakout—(Slang) to unscrew or disconnect pipe and/or drill collars.

Bridge—Caving in of a wellbore, causing an obstruction in the well.

Bullseye—A well drilled with no deviation.

Button bit—Tungsten carbide bit.

Cable method—A method for extinguishing a fire without the use of heavy equipment, used by Wild Bunch Hellfighters.

Cable tool rig—An older type rig that is now used only rarely. It uses a cable and a solid drill bit to actually pound a hole in the ground.

Casers—Persons with special tools for running casing into the hole.

Casing—Pipe used in a well.

Cathead—An extension of the drawworks drum used to lift heavy objects at the rig.

Catline—Rope attached to a cable used on the cathead to lift heavy objects.

Catwalk—Ramp used to get pipe and other heavy equipment to the rig floor.

Cellar—Space left between the rig floor and the ground to allow the installation of the bradenhead at ground level (point of drilling).

Cement job—Placement of cement around the annulus or casing by pumping a cement slurry down through the casing.

Cement manifold—A set of valves used to close off a well after cementing and to pump the plug after the cement is pumped.

Cement slurry—A mixture of cement pumped into the well to cement the casing or to set the plug.

Cementer—Person in charge of all cement jobs. He directs the mixing and pumping of the cement downhole.

Centralizer—Device attached to the collars of casing to center the casing in the hole.

Chaining out the hole—To use a breakout chain to breakout drill pipe while coming out of the hole. This keeps the drill stem from rotating in the open hole while tripping pipe.

Choke line—Piping from the BOP that allows annulus fluid to be diverted to a system of valves so gas and oil kicks can be controlled.

Circulate and weight—A method of killing a kick used widely on the Gulf Coast.

Company man—Person who works directly for the operator or the independent consultant on the well.

Conductor pipe—Short pipe used to keep the surface sands from sloughing in the wellbore. It is usually driven in the hole by a diesel hammer. A conductor in swampy areas and on the Gulf Coast is permanent and in hard rock areas is temporary.

Consultant—The top hand in the field, this person works for himself on a daily rate and is in charge of the drilling operation in the field. Consultants usually drill wildcat wells, because most companies do not have men trained to drill in unknown areas.

Coring—Obtaining part of the downhole geology, intact, through the use of a core barrel.

Crooked hole country—Geology that is faulted and fissured, making it hard to drill a straight hole. (Drill bits follow the path of least resistance through formations.)

Crown—Top of the derrick where a system of sheaves is used to handle the drill line.

Crownamatic—An automatic cut-off device that stops the blocks from going through the crown while you are tripping pipe.

Curve—The hardest part of a horizontal well; it turns the string and allows it to bend 90°.

Cuttings—Pieces of rock formation, cut by the drill bit; drilling fluids circulate cuttings to the surface.

Daywork—A price rate in which a contractor charges for the rig on a daily rate instead of by footage drilled.

Deadline— End of the drilling line attached to the derrick or substructure.

Deadman—Anchor system to handle guy-wires to steady the derrick.

Degasser—A device used to expel the gas from the drilling mud while drilling or circulating.

Delay rental—Payment from an operator to a landowner for the privilege of delaying drilling; penalty clause in the lease contract.

Density—The weight per gallon of a given volume of fluid compared to the weight per gallon of water. The base density of fresh water is 8.33 lb/gal.

Derrickman—Person in charge of mixing mud, maintaining the mud pumps, and stacking pipe in the fingerboard when tripping.

Dog house—Sometimes called the change house, it is a place for the crew to change clothes and store personal items.

Dog leg—A deviation in drilling that can be caused by a number of things. It is a serious condition that could lead to the drill pipe stacking.

Double rig—Rig that allows for two joints of pipe to be stacked in the derrick during tripping; as opposed to a triple rig, which allows for three joints. The joint set is called the stand.

Drawworks—The machinery that moves the traveling blocks up and down the derrick by using a cable attached to the drawworks drum.

Drill bit—The device that cuts the formation and allows the hole to be drilled.

Drill collars—Heavily walled pipe used to give weight to the bit and to help keep the hole straight by causing a pendulum effect on the drill stem.

Drill pipe—Pipe above the drill collars used to connect the bit with the surface.

Drill stem—Drill pipe and drill collars with the bottom hole assembly attached.

Drill stem test (DST)—A test to check wellbore pressure and the fluids at certain depths; used to make a temporary completion to check for oil or gas in commercial quantities.

Driller—Operator of the rig and boss of the rig hands, hiring and firing the crew working under him.

Drilling break—Time when penetration rate increases as a result of changes downhole (abnormal pressure, sand, etc.).

Drilling contract—A contract between the drilling contractor and the operator for drilling a well, which spells out all responsibilities and liabilities of both parties.

Drilling contractor—Company that owns the rig and employs the rig crew.

Drilling jar—A tool used to move the drill pipe up or down when it is stuck.

Drilling mud—Material added to water to form the fluid used to drill a well.

Drilling spool—A spacer in the BOP stack.

Dry hole—A well that produces no commercial quantities of oil or gas.

DST—See *Drill stem test.*

Elevators—A latch to attach drill pipe of collars to the traveling blocks by bales.

Engineer—Person who writes the drilling program and sets up the AFE.

Equivalent circulating density (ECD)—The effective mud density at a specified depth when circulating. When mud is pumped it is actually heavier than its actual weight due to pressure. For example, 10.2 lb/gal mud when pumped at certain pressure would have an ECD of 10.4 lb/gal.

Equivalent mud weight (EMW)—The effective mud density at a specified depth when pressure is imposed on the top of a mud column. It is used to test a casing shoe; by adding pressure the mud can be brought to a higher test weight downhole.

Fast line—The end of the drilling line attached to the drum on the drawworks.

Filling the hole—To pump fluid through the fill line, keeping the hole full of fluid to eliminate blowouts.

Fingerboard—A rack located in the derrick for stacking the stands of pipe while tripping.

Fish—An unattached pipe or object in the hole (lost object in the wellbore).

Fishing job—To bring a lost or unattached object out of the wellbore.

Fishing tools—Tools used to obtain objects left in the well.

Flange up—(Slang) to finish a job; the final connection of pipe; also means to quit and leave your job.

Flare line—A line coming from the gas buster to flare gas while the well is being drilled or flowed.

Float equipment—A one-way valve for holding cement in place after it has been pumped.

Flowing well—A well requiring no pumping to bring hydrocarbons to the surface.

Fluid loss chemical—A chemical used to tighten the wellbore and keep water from going into the formation as fast.

Footage rate—A price rate in which the drilling contractor charges a price based on footage drilled.

Frac tanks—Tanks used to handle oil on location when the well is flowing.

Gel—Substance used to give viscosity to drilling fluids. See also *Bentonite*.

Geologist—Person who evaluates rocks and core samples, mud logger reports, and drilling logs.

Geolograph—A device used to record the weight of the string, trip time, downtime, footage, time per foot, etc.

Going flat—(Slang) going in a horizontal direction with the drill bit.

Going in the hole—Running pipe into the wellbore.

Green cement—Cement that has not yet dried.

Guide shoe—An insert attached to the bottom of the first joint of casing to guide the string away from downhole obstacles.

Heavy wate drill pipe—Reinforced drill pipe similar to collars that is flexible and can be used in trouble spots such as dog legs.

Hole deviation—The wellbore variation from vertical.

Hole sloughing—When the formation falls apart and enters the wellbore; usually causes a bridge.

Hook load—The weight of the drill stem.

Horizontal drilling—A method to drill a zone and stay in the pay zone for a large distance; in chalk wells the drill string can drill through multiple oil zones.

H₂S inhibitor—Chemicals that neutralize the invasion of hydrogen sulfide gas into the drill stem.

Hydrogen sulfide (H₂S)—Poisonous gas sometimes encountered in drilling.

Inclinometer—A device that reports drift in the well.

Inflatable packer—Used to finish a horizontal well, usually set about 50 ft above the intermediate string.

Intermediate string—Casing set in the wellbore to continue drilling deeper. Used to keep the wellbore from sloughing and when high-pressure gas is expected.

Inverted blowout interceptor tool—Tool used to kill an open-hole blowout, used by Wild Bunch Hellfighters.

Jet nozzle bit—A drilling bit that utilizes a jet insert to control hydraulics.

Jetting out—Cleaning out the mud pits. Solid particles are jetted into the reservoir through bottom suction.

Junk—Used equipment or parts.

Junk basket—Container for holding junk or used parts or for storing unused equipment.

Kelly—Square, hexagonal, or triangular pipe at the top of the drill string used to rotate the string when it is set in the kelly bushing.

Kelly bushing (KB)—Device on rotary table that allows the kelly to turn and slip vertically while drilling operations are underway.

Keyseat—A groove worn in the side of the wellbore by the drill stem. Pipe can become stuck while moving through the keyseat. Keyseats sometimes form when excessive circulation and rotation are needed due to gas kicks. A short trip normally keeps the keyseats out of the hole.

Keyseat wiper—Short sub used to enlarge the hole by using blades larger than the drill collars. Stabilizers and roller reamers do a better job than the wiper.

Kick pad—A field modification welded to motors to create more of an angle of attack in making the curve.

Kill line—A pipe line to bypass the kelly, if needed, to control the well; hooked to the BOPs.

Knowledge box—Box on the floor where the driller keeps records, notes, scratch pads, etc.

Laminar flow—A type of flow that tumbles the cuttings in the drilling fluid and causes cuttings to take longer to reach the surface.

Latching on—To attach elevators to drill pipe or drill collars.

Laydown machine—A machine that picks up or lays down drill pipe and casing through the V-doors.

LCM—See *Lost circulation material.*

Lease—See *Oil and gas lease.*

Liner casing—String of casing set below the surface.

Liner hanger—Device used to land a liner on preexisting casing.

Logging the well—To record data on formations below the surface; to evaluate pay zones through electrical or radiation tools.

Lost circulation—When drilling fluid enters a porosity zone in a formation and the formation takes fluid.

Lost circulation material (LCM)—Substance used to seal a zone losing fluid.

Making up a joint—To add a section or joint of drill pipe to the drill stem to continue drilling.

Making hole—Drilling ahead or deeper, "turning to the right."

Making a trip (tripping)—To pull drill pipe to change bits or tools and returning the drill pipe to the hole with the new bit or tool.

Marsh funnel—Testing funnel for evaluating viscosity of drilling fluids, primarily for field tests.

Measurements-while-drilling (MWD) tools—Tools that take the measurements and allow the operator to know the direction of horizontal or any directional tools.

Mixing mud—To add barite, gel, or other chemicals to the drilling mud.

Monkey board—Platform near the top of the derrick where the derrickman works during trips.

Motorman—Person in charge of preventive maintenance for rig engines.

Mousehole—Hole next to the rotary used for placing drill pipe before it is connected to the kelly as joints are added.

Mud—Drilling fluid used in rotary drilling to maintain borehole pressure equilibrium and to remove bit tailings from bottomhole.

Mud balance—Tool to measure the weight of mud in pounds per gallon.

Mud engineer—Person in charge of the drilling fluids on location. One is assigned to each location.

Mud engineering—The science of drilling fluids.

Mud hopper—A container used on location to keep large quantities of barite.

Mud logger—Person in charge of monitoring the mud samples.

Mud logging unit—A trailer set up on location that contains devices for evaluating the mud coming over the shale shaker.

Mud plug—Heavy mud slurry sent down the hole to plug off a high pressure zone (slang—mud pill).

Mud pump—Duplex or triplex pump used to pump mud down the drill stem to circulate the drilling mud.

Mud tanks—Tanks for conditioning and cleaning returned mud.

Nipple up and nipple down—(Slang) to put together or take apart.

Nitrogen bomb nozzle—Device used by Wild Bunch Hellfighters to direct a flow of nitrogen on a fire or to supercool a well.

Oil and gas lease—Legal document from a landowner giving the operator the right to explore for and produce oil and gas on the landowner's property.

Open hole—An uncased wellbore.

Operator—The oil company responsible for drilling and producing an oil or gas well.

PDC bits—Bits used to drill the horizontal portion of the well and on directional wells. They normally will drill four to five wells.

Penetration rate—Rate in feet per hour at which the bit drills the open hole formation; it is the major cost factor in drilling economics.

Pins and boxes—Refers to the male (pins) and female (boxes) connections.

Pipe dope—The substance used to lubricate and help seal drill pipe and all connections in the drill stem; usually made of lead or magnesium compounds.

Produce-while-drilling (PWD) equipment—The equipment used to produce oil for sale during drilling of the well.

It also allows the drilling fluid to be returned to the mud tanks and the gas to be flared.

Racking pipe—To place pipe in the derrick while tripping.

Rams—Device in BOPs for sealing off the hole. There are four types: pipe rams, casing rams, blind rams, and shear rams.

Rat hole—A hole in the rig floor used to store the kelly and swivel while making a trip; also refers to the portion of the hole drilled deeper than the pay zone to allow for mistakes.

Reaming the hole—To enlarge the wellbore and clean out the hole to eliminate tight holes.

Reverse circulation—To pump down the backside (annulus) and back up the drill stem (used for squeeze jobs).

Rig—Moveable equipment used to drill wellbores; a rig consists of many parts including mud tanks, pumps, rig floor, dog houses, derrick, drawworks, engines, generators, etc.

Rig hydraulics—The circulation of drilling fluids.

Rig up—To get ready to drill the hole.

Rotary kelly bushing (RKB)—Attaches the kelly to the rotary table. It transfers the twist produced by the table to the kelly.

Rotary rig—A rig that rotates the pipe by using a rotary table which in turn drills a hole.

Rotating head—Device that allows the operator to drill well under pressure while rotating. Without this device the horizontal well could not be drilled.

Roughneck—Worker on the rig.

Roustabout—Oilfield laborer.

Royalty—A percentage of all oil and gas produced, paid to the landowner or other recipient, free of all expenses.

Service hand—A specialty company's employee who performs a special service for the operation.

Setting casing—To run and cement casing in the wellbore.

Shale shaker—A service used to separate the cuttings from the mud as the mud comes to the surface.

Shear rams—Hydraulic closures used to close and seal the hole in an emergency; closing the shear rams will cut the drill pipe while sealing the hole.

Shock sub—A shock absorber used above the drill bit to reduce bouncing of the drill stem.

Shoe—See *Guide shoe*.

Short trip—Pulling several stands of pipe in order to clean the bottom of the hole.

Shut in—To shut a well in by closing the annular, pipe, or blind rams.

Side tracking—Drilling around a fish left in the hole.

Slips—Wedges used to hold the pipe vertically in the rotary tables.

Slug the pipe—To put a heavy slug of drilling fluid into the drill stem to create more hydrostatic pressure. The slug pushes fluid down the pipe so you don't have to pull a wet string out of the hole. This is also called pumping the slug.

Slush pumps—Mud pumps.

Spinning chain—A chain used to make up or break out drill pipe.

Spread the kick—A method to kill a horizontal well with fresh water. It keeps heavy brine water from being pumped into the well.

Squeeze jobs—Also called squeeze cementing, this involves forcing cement into the wellbore to seal off casing seats and to seal off weak formations.

Stabbing board—Device used to stab casing when running in the hole.

Stabilizer—A sub with fins the same size as the bit used to stabilize drill collars in the wellbore.

Stacked rig—A rig with no work.

Stand of pipe—Either double or triple joints of pipe made

up in a derrick. The number of joints used depends on the height of the derrick.

Stand pipe—Permanent pipe used to connect the rotary hose for mud flow.

Sub—Device used to join two pipes of different sizes together.

Substructure—Structure that the derrick, drawworks, and motors sit on.

Superchoke—A device used to control wells that have "kicked" and which keeps sand from cutting the main valve so much during a blow.

Surface casing—The first string of casing set in the well. It is used to isolate water sands to keep them from becoming contaminated.

Swabbing—A procedure for applying suction within the casing or tubing to draw fluid from the reservoir.

Tag bottom—Touch bottom.

Tankers—Trucks used to take oil off location.

TD—See *Total depth*.

Tearing down—To get ready to move the rig.

Tight hole—A well that is drilled in secret, with no data going out concerning the drilling operation. Unless the tight hole is in the middle of nowhere, everyone still knows what is going on through oilfield gossip.

Tongs—A large pipe wrench suspended from the derrick used to make up or breakout a drill stem.

Toolpusher—The boss of the rig.

Torquing up—Stress placed on the rotary due to the bit or drill string dragging or hanging up.

Total depth (TD)—Depth to which the well is to be drilled.

Traveling block—The block and tackle used to pull the drill stem; it moves up and down the derrick to trip the pipe and perform drilling functions.

Triple rig—Rig that will allow for drill pipe to be stacked three joints at a time.

Tripping—The going in or coming out of the hole with drill pipe, drill collars, casing, or tubing

Turbulent flow—Drilling fluid flow that allows cuttings to come to the surface without tumbling and falling back.

Turn to the right—An expression indicating that everything is going well.

Twist-off—Pipe failure downhole due to pipe fatigue or mishandling; also slang for quitting an oilfield job.

U.S.G.S.—United States Geological Survey.

V-door—Entry on the front of the rig to supply tools, mainly drill pipe and collars and casing.

Visbestos—Asbestos material used to increase viscosity and push cuttings to the surface. It requires careful handling as asbestos is a health hazard.

Viscosity—Measure of fluids' resistance to flow.

Wait-and-weight—A method used to kill a well by first weighting up the mud to the kill weight, then pumping it down the hole.

Wait on cement (WOC)—To wait for the cement to dry before resuming operations.

Wash to bottom—The lowering of drill pipe to the bottom of the hole with the pumps on but without the pipe turning.

Water cushion—Water run inside a pipe to keep the pipe from collapsing while a drill stem test (DST) is run.

Wear ring—A ring installed in the bradenhead to keep the kelly from wearing out the bradenhead.

Wedding band—A safety clamp used on nonindented drill collars.

Wellhead—A machined metal product that is welded or screwed on the surface casing and allows the BOP stack to be bolted to it, making the well safe. When the well is completed the Christmas tree is bolted to it.

Index

Abstract company, 2, 333
Abstract of title, 2, 7, 333
AFE (*see* Authorized field
    expenditures)
Agreed footage, 30
Annular capacity of tubular products
    in a wellbore, calculations for,
    307
Annular preventer, 76, 97, 101,
    219, 221–223, 238, 251, 333
Annular volume and height,
    calculations for, 315–318
Authorized field expenditures (AFE),
    333
    intangible items, 13–25
    tangible items, 25–26
Automatic driller, 134, 136, 333
Automatic fill float collar, 71

Barite, 48, 121
Baseline, 8
BHA (*see* Bottom hole assembly)
Bit, 60, 334
    box, 113
    grading wear on, 114
    jets, 113, 334
    life of, 112–113
    program, 43
    pulling the, 114–116
    sub, 85, 334
    types of, 111–112
        diamond, 113
Bit-to-surface time, 312
Black Magic™, 167–170

Blind ram, 76, 83, 333
Blowout preventer (BOP):
    nippling down, 75–83, 127, 238
    nippling up, 75–83, 127, 203
    pressure testing sheet, 321
    systems, 76, 125, 135–136
    testing, 82, 83
Blowouts, 76, 140, 333
    calculations for control of, 247–
        274
Bonus, 2, 4, 333
BOP (*see* Blowout preventer)
Bottom hole assembly, 50, 159, 333
    checklist, 321
    inspecting, 136
    purpose of, 91
    setting up, 84–96
Bottom hole pressure, calculations
    for, 306
Bradenhead, 50, 75, 76, 333
Bridging, 61, 138, 153, 191, 267,
    333
Bumper jar, 162
Bumping the plug, 73
Button bit, 112, 333

Cable tool rig, 46, 333
Calculations:
    annular capacity of tubular
        products in a wellbore, 307
    annular volume and height, 315–
        318
    bit-to-surface, 312
    blowout, 247–265

Calculations (*Cont.*):
  bottom hole pressure, 306
  bottoms up, 231
  capacity, 306–307, 314–315
  casing length, 194
  cement, 308–309
  common oilfield, 305–313
  controlling hole deviations, 143–147
  drill collars, 309–310, 323
  equivalent mud weight, 66, 97, 99, 305, 310–311
  kick control, 247–265
  kill mud, 313
  plug-and-abandon, 243–246
  pump stroke, 312
  slurry, 308–309
  slurry density, 66–68
  slurry yield, 66–68
  surface-to-bit, 311–312
Capacity, calculations for, 314–315
Capacity of tubular products, calculations for, 306–307
Carried interest, 3
Casers, 21, 62, 124, 242, 333
Casing, 60–65, 75, 97, 194–195, 201, 229–232, 242, 244–246, 333
  installing, 61–65, 194–197, 229–234
  leaks in, 97
  program, 43
Cellar, 58, 59
Cement basket, 64, 65
Cement book, 314–318
Cement engineers, 43, 64
Cement manifold, 61, 64, 71, 106, 333
Cement plug, 143
Cement program, 43

Cementers, 125, 333
Cementing, 3, 20, 66–74, 195–197, 232–237
  calculations, 66–68, 195–196
  problems, 68, 275
  program, 43
  types of, 64, 195
Centralizer, 62, 63, 125, 276, 333
Changeover sub, 89
Choke, 56, 127–128, 215–216, 218–221, 223, 251, 263
  line, 333
Choke manifold system, 48
Circulate-and-wait method, 254, 282, 333
Clean-up, 15
Collar weight measurement, 91–94
Collars (*see* Drill collars)
Commencement date, 30
Companyman's shack, 48
Completion of wells, 3
Conductor casing, 25
Conductor pipe, 57, 333
Consultants (*see* Drilling consultants)
Contract drilling:
  items of, 3, 28–34
  types of, 28–29
    IADC footage drilling, 279–304
Contractors, 32
Controlling hole deviations, calculations for, 143–147
Coring heads, 184
Coring the well, 184–188, 333
  analysis and, 20
  sample worksheet, 322
Crooked hole country, 87, 333
Crownamatic, 134, 135, 333

Daily reports, 240–241, 320
Daywork contract, 28–34, 333

Degasser, 47, 48, 50, 128, 129, 131, 333
Delay rental, 2, 4, 333
Depth of well, 30
Derrickman, 49, 250, 333
  finishing the well, 238
  kicks and, 250, 253
  mud hopper and, 49
Development well, 1
Deviation chart, 144
Diamond bit, 113
Diamond core head, 184
Differential stuck, 88, 166–171
Difficult formations, 32
Directional driller, 16, 42, 54, 94, 96, 204, 206–208, 209, 211, 212, 222
Directional drilling, 6, 84, 94
Dog legs, 143, 148, 149, 185, 333
Downtime, 31, 62, 152–153
Drift chart worksheet, 326
Drill bits, 43, 60, 85, 87, 111–117, 333
Drill collars, 60, 85, 87–94, 148, 163, 167, 333
  calculations, 309, 310, 323
  displacement tables, 328–329
  washouts, 148, 155, 170
Drill pipe, 61, 136
  See also pipe
Drill stem test (DST), 20, 175–183
  information yielded, 179
  problems with, 175, 182–183
  tools, 154, 175–183
  water cushion, 183
  worksheet, 322
Drilling, 60, 124–126
  ahead, 124–126
  bits, 18, 111–117, 152, 155
  contracts, 3, 28–34, 279–304, 333

Drilling (*Cont.*):
  directional, 6, 16, 84, 94
  equipment:
    annular preventer, 39, 40, 76, 78–80, 82, 216, 219, 221–223, 251
    automatic driller, 134, 136
    blowout preventer, 75–83, 134–136, 238, 321
    See also Blowout preventer
    bottom hole assembly, 50, 84–96, 136, 159, 321–322
    See also Bottom hole assembly
    bradenhead, 75, 83
    casing, 38, 39, 61, 64, 72, 73, 194, 229–231
    See also Casing
    casing head, 53, 75, 76
    changeover sub, 89
    collars, 60, 85, 87–94, 148, 163, 167, 333
    conductor pipe, 57
    coring heads, 184
    crownamatic, 134–135
    degasser, 47, 48, 50, 128, 129, 131, 333
    drilling line, 134–135
    DST (drill stem test) tools, 154, 175–183
    failure, 152–153
    fishing tools, 161–174
    float collar, 71, 97, 125
    float equipment, 21
    geolograph, 134, 136, 241, 320
    guide shoe, 62, 97, 125
    jar, 47, 50, 52, 88, 89, 162
    kelly bushing, 61
    laydown machine, 21, 231, 245
    line, 134–135

Drilling, equipment (*Cont.*):
    liner, 25, 197–202
    logging tools, 189–193
    mud hopper, 47, 48
    pipe, 136, 155–157, 170, 245
    pipe rack, 61
    pumps, 134–135
    rental equipment, 19, 47–56,
        127–133, 239, 325
    rig, 45, 46
    safety clamp, 88, 89
    shale shaker, 47, 48, 60, 118,
        140
    shock sub, 51, 84–85, 114
    squeeze tool, 97, 101, 105
    stabilizer, 47, 50, 85
    superchoke, 47, 251
    wear ring, 50
  on Gulf Coast, 29, 57, 86, 97,
      99, 109, 112, 124, 182
  in hard rock, 57
  in Louisiana, 14, 21
  mud (*See* Mud)
  personnel:
    caser, 21, 61–62, 124, 242
    cementer, 125
    derrickman, 49, 238, 250, 253
    drilling consultant (*See* Drilling
      consultant)
    drilling contractor (*See* Drilling
      contractor)
    drilling engineer (*See* Drilling
      engineer)
    geologist, 1, 3, 22, 175, 176
    motorman, 253
    mud engineer, 43, 48, 120,
      122, 139
    mud logger, 20, 49, 240
    toolpusher, 43, 48, 113, 134,
      152–153, 239, 253
  problems, vertical, 138–160

Drilling, (*Cont.*):
  problems, horizontal, 214–226
    fishing, 161–174, 182–183,
        191–192
    keyseat, 116, 143, 148, 152
    lost circulation, 139–143
    sticking and torquing pipe,
        148–151
    twist-off, 155, 157–159
  procedure, 36, 38
  prognosis, 35–46
  regulating agency, 1
  slant, 6
  Texas Railroad Commission, 1
  transactions, 1
Drilling consultant, 333
  arrival at rig site, 57
  engineer and, 147
  hiring of, 1, 29, 35–36
  responsibilities, 3, 47, 111
Drilling contractor, 3, 33–34, 134
  fees of, 15
Drilling engineer, 63, 82, 147,
  333
Drilling jars, 47, 50, 52, 88, 89,
  162
  charts, 90–93
Drilling rig, 45–46
Drive hammer crew, 57
Dropping a cone, 115–116
Dry hole (duster), 242, 333

Electric logging, 20
Engineer, 63, 82, 147, 333
Equipment failure, 151
Equivalent circulating density, 139,
  333
Equivalent mud weight, 66, 97, 333
  example calculation, 99, 101, 311
Exhibit A, 33

Fishing job, 3, 161–174, 182–183,
191–192
tools, 161–174
Float collar, 71, 97, 125
Float equipment, 21
Flow shows, 51, 320
reports, 241
Fluid loss, 121
Footage contracts, 28, 34, 279–304,
333
Formation invasion, 120
Free point, 167, 170
Fuel, 19

Gas indicators, 51
Gas kicks, 51
Gel, 61, 121
Geologist, 1, 3, 22, 175, 176
DST and, 175–176
Geolograph, 134, 136, 241, 320,
333
Guide shoe, 62, 97, 125, 333
Gulf Coast drilling, 29, 57, 86, 97,
99, 109, 112, 124, 182
Gyro survey, 147

Hard rock area drilling, 57
Hole:
bridging, 61–62, 153–155, 191,
267
chaining out, 62
depth of, 64
deviation, 143
pebbles, 61
sloughing, 62
straight, 61
Hydrogen sulfide, 120, 333
Hydrolic wrench, 125
Hydrolics, 113

IADC (*See* International Association
of Drilling Contractors)

Independents, 1, 2, 35
Insurance, 16, 22
Intercoms, 47, 48
Interest, in wells, 3, 5
Intermediate casing, 25, 242
Intermediate string, 194–197, 333
International Association of Drilling
Contractors (IADC), 29, 33
footage drilling contract, 279–304
Investors, 3, 5, 36, 215

Jars, 47, 50, 52, 88, 89, 162

Kelly, 50, 333
Kelly bushing, 61, 333
Key maintenance, 134–135
Keyseat, 116, 143, 148, 152, 333
Kicks, 78, 113, 127, 131, 135, 204,
210, 215–225
controlling, 247–265
kill methods, 254–265
calculations, 258, 260–262,
313
stations of personnel, 253
signs of, 250
Kill line, 76–80
Kill mud calculations, 313

Land:
damage to, 14
delineation of, 7–11
ownership of, 6
Landman, 2, 4
Landowners, 6, 14
payments to, 4–5
rights of, 2, 4, 6
Laydown machine, 21, 231, 232,
245, 333
Leaks, 75, 136, 214, 218–219, 222,
224
Lease, oil and gas, 2, 5, 333
Liner casing, 25, 333
hanging, 197–202

Logging, 3, 20, 189–193
  tools, 189–193
Long string, 229–237
Lost circulation, 51, 121, 218, 333
Lost equipment, 32, 33
Louisiana drilling, 14, 21

Major company, 2, 35
Metes and bounds description, 8–11
Mineral rights, 6
Mobile phone, 47, 53
Monkey board, 49, 333
Motorman, 253, 333
Mud, 47, 48, 118–123, 333
  equivalent mud weight, 66, 97,
      99, 101, 310–311
  flow, 47, 51
  makeup, 121
  report, 122
  types of, 122
Mud engineer, 43, 48, 120, 121,
    139, 333
Mud hopper, 47, 48, 333
Mud logger, 20, 47, 49, 240, 333
Mud pumps, 118, 122, 333
Mud tanks, 57, 118, 333
Multiple completions, 3
Multishot survey, 147

Nippling up, 76, 203, 324, 333

Offset well, 6
Oil and gas, rules governing, 6
Oil and gas drilling transaction, 1–
    10
Oil and gas lease, 2, 4, 333
Oil company (*See* Operator)
Oil well (*See* Well)
Oilfield terminology, 35
One-inch job, 68
Operator, 1, 35, 333
  personnel needs, 2, 3

Operator (*Cont.*):
  responsibilities, 3, 31–32, 49, 101

Paperwork, 238–241, 253
Penetration rate, 115, 136, 152,
    250, 333
Permeability, 189
Personnel (*See* Drilling, personnel)
Petroleum engineer, 13
Petroleum land titles, 6
Pipe, 54, 94, 136, 155, 157, 177,
    186
  strapping, 61, 229
  washout, 11, 155–160
Pipe dope, 159, 333
Pipe rack, 48
Pipe ram, 76, 83, 333
Pit gain, 51
Pit indicator, 51, 320
Plug, 73, 234–235, 245
Plug-and-abandon, 3, 26
  calculations, 243–245
  procedures, 243–245
Pooling, 6
Pressure testing, 82
Primary term, 4
Problems, in drilling horizontal,
    214–226
Problems, in drilling vertical, 138–
    160
Production of well, 3
Prognosis, 35–46, 97, 99, 120, 122,
    194
  items of, 36
Pump output table, 331–332
Pump stroke calculations, 312
Pumper, 3
Pumps, 134–135

Rams, 76, 333
Ranges, 8
Rat hole, 234, 333

Rate of penetration, 115, 136, 152, 250, 333
Record keeping, 95
Rectangular survey, 7–11
Reimbursable costs, 31
Rental items, 16, 47–56, 238, 325
Repairs, 31
Reports, 239
Rig, 45–46, 333
Rig hands, 57
Rig move, 16
Rig up, 57
Roller-and-cutter head, 184
Rotary rig, 45
Rotary table, 45, 88, 116
Royalties, 2, 4–5, 333

Safety clamp, 88–89
Sections, of rectangular survey, 8
Service companies, 124 126
  acidizing, 3
  casing, 3
  cementing (*See* Cementing)
  drive hammer, 57
  fishing, 3, 161–174, 182–183, 191–192
  fracturing, 3
  geologist, 1, 3, 22, 175
  logging, 3, 17, 189–193
  mud (*See* Mud)
  perforating, 3
  scheduling of, 231–232
  spud unit, 57
Shale shakers, 47, 48, 60, 118, 140, 333
Shock sub, 51, 84–85, 114, 333
Sidewall core gun, 185
Site preparation, 14
Slant drilling, 6
Slurry calculations, 308–309
Slurry density formula, 67
Slurry yield formula, 67

Spud in, 57
Spud unit, 57
Squeeze jobs, 21, 101, 333
  procedures, 101–110
  successful, 101
  tools, 97, 101, 103, 105
  unsuccessful, 109
Stabilizer, 47, 50, 85–86, 333
Standard float collar, 71
Standby rate, 31
Stop ring, 62
Superchoke, 47–49, 127, 128, 216, 218–221, 251, 263
Surface bit, 111
Surface casing, 5, 25, 57, 61, 71–73, 75, 333
  drilling out, 97
  worksheet, 325–326
Surface damage, 5
Surface rights, 6
Surface shoe, 99
Surface-to-bit calculations, 311–312
Surveying, 13

Temporary conductor, 39
Tests, 101, 189–190
Texas Railroad Commission, 1
Thread cleaners, 125
Three-ram system, 76
Titles, land, 2, 6–11
Tongs, 62, 333
Toolpusher, 43, 48, 113, 134, 152–153, 239, 253, 333
Townships, 8
Transactions, oil and gas, 1–10
Trucking and hauling, 20–21
Tungsten carbide bit, 112
Twist-off, 155–160, 333
Two-ram system, 78, 81

Unitization, 6

V-door, 61, 333
Visbestos, 61, 182, 333

Wait-and-weight method, 254–262, 333
Wash pipe, 170–171
Washouts, 152
Wear ring, 53, 333
Weight on bit, 43, 92
Welder, 57, 75
Well:
  completion, 3, 275–278
  depth, 30
  development, 1
  drilling, 1, 35
  finishing, 238–241

Well (*Cont.*):
  interests in, 3
  location, 14, 30
  offset, 6
  sign requirements, 1
  successful, 4
  turnkey, 13
  wildcat, 1, 122
Well site, 19, 22
Wellcat well, 1, 122
Wellhead, 25, 333
Wireline coring, 184–188
Wireline logs, 124
Wireline survey, 60, 143
Work stoppage rate, 31
Working interest, 5

About the Author

Byron "Duke" Davenport is a consultant to the oil and gas drilling industry and is president and owner of Davenport Horizontal Drilling Consultants, which furnishes well-site supervision for vertical and horizontal wells. In addition to running his consulting firm, Dr. Davenport heads the Wild Bunch Hellfighters in extinguishing oil well fires using nitrogen technology. He participated in efforts to extinguish the Kuwait oil well fires, furnishing nitrogen technology.

Dr. Davenport holds a Ph.D. in petroleum engineering and is a member of The Society of Petroleum Engineers.